Jochen Werner

Optimization
Theory and Applications

Advanced Lectures in Mathematics

Edited by Gerd Fischer

Jochen Werner
Optimization. Theory and Applications

Manfred Denker
Nonparametric Statistics

Jochen Werner

Optimization Theory and Applications

Friedr. Vieweg & Sohn Braunschweig/Wiesbaden

CIP-Kurztitelaufnahme der Deutschen Bibliothek

Werner, Jochen:
Optimization – theory and applications / Jochen Werner. – Braunschweig; Wiesbaden: Vieweg, 1984.
(Advances lectures in mathematics)
ISBN 3-528-08594-0

1984

Produced by IVD, Walluf b. Wiesbaden
Printed in Germany

ISBN 3-528-08594-0

PREFACE

This book is a slightly augmented version of a set of lectures on optimization which I held at the University of Göttingen in the winter semester 1983/84. The lectures were intended to give an introduction to the foundations and an impression of the applications of optimization theory. Since infinite dimensional problems were also to be treated and one could only assume a minimal knowledge of functional analysis, the necessary tools from functional analysis were almost completely developed during the course of the semester. The most important aspects of the course are the duality theory for convex programming and necessary optimality conditions for nonlinear optimization problems; here we strive to make the geometric background particularly clear. For lack of time and space we were not able to go into several important problems in optimization - e.g. vector optimization, geometric programming and stability theory.

I am very grateful to various people for their help in producing this text. R. Schaback encouraged me to publish my lectures and put me in touch with the Vieweg-Verlag. W. Brübach and O. Herbst proofread the manuscript; the latter also produced the drawings and assembled the index. I am indebted to W. Lück for valuable suggestions for improvement. I am also particularly grateful to R. Switzer, who translated the German text into English. Finally I wish to thank Frau P. Trapp for her care and patience in typing the final version.

Göttingen, June 1984 Jochen Werner

CONTENTS

Bei dem studio der Mathematik kann wohl nichts stärkeren Trost bei Unverständlichkeiten gewähren, als daß es sehr viel schwerer ist eines anderen Meditata zu verstehen, als selbst zu meditieren.

G. Chr. Lichtenberg

§ 1 INTRODUCTION, EXAMPLES, SURVEY

An optimization problem consists in minimizing a function $f : M \to \mathbb{R}$ on a given set M. $\overline{x} \in M$ is a solution of

(P) minimize $f(x)$ on M,

if $f(\overline{x}) \leq f(x)$ for all $x \in M$. f is often called cost or objective function, M the set of feasible solutions. The value of the optimization problem (P) is

$$\inf (P) := \begin{cases} \inf\{f(x) : x \in M\} & \text{if } M \neq \emptyset \\ + \infty & \text{if } M = \emptyset. \end{cases}$$

If (P) has a solution, then we shall always write min (P) instead of inf (P). The maximization of a function $g : M \to \mathbb{R}$ on M can of course be reduced to the problem (P) by introducing $f := - g$.

In this introduction we shall begin by giving several examples of optimization problems. We shall see that widely differing types of problems can be formulated in our general setting. At the same time we hope to whet the reader's appetite for concrete problems.

1.1 Optimization problems in elementary geometry

According to CANTOR [12, p. 228] the first optimization problem in the history of mathematics occurs in EUCLID's Elements, Book VI, Theorem 27. The problem is essentially the following:

1) Find a point E on the side BC of the triangle ABC such that the parallelogram ADEF with vertices D resp. F lying on the sides AB resp. AC has maximal area.

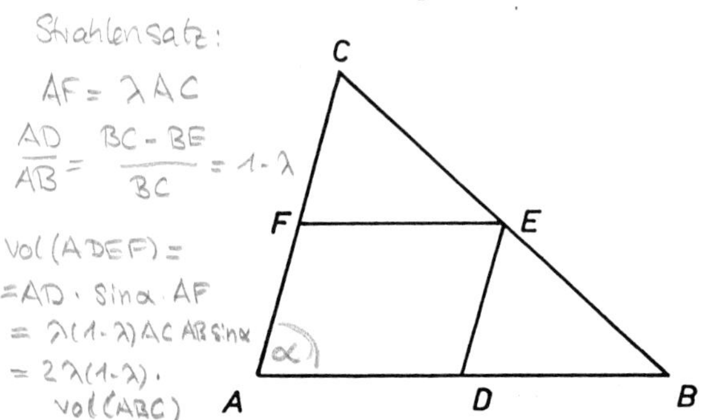

The solution is obviously given by choosing E to be the midpoint of BC. In fact for arbitrary E on BC with $\lambda := BE/BC$ the area of the corresponding parallelogram is $\text{area}(ADEF) = 2\lambda(1-\lambda)\text{area}(ABC)$ and this function is maximal for $\lambda = 1/2$.

HERON (ca. 100 BC) gave a solution to the following problem:

2) On a given line find a point C such that the sum of the distances to the points A and B is minimal.

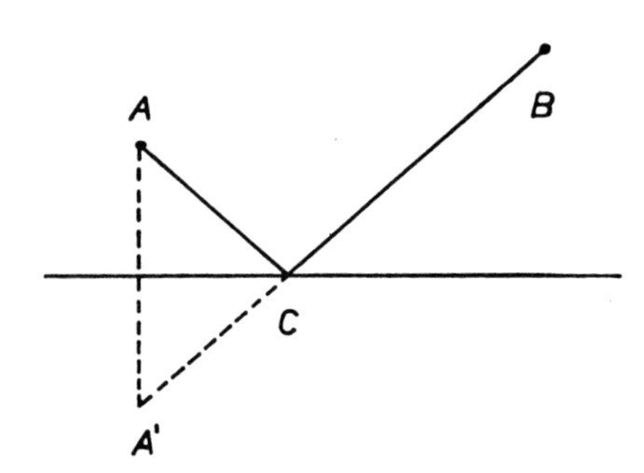

If A and B lie on opposite sides of the given line, then obviously the intersection of the line with the segment AB is the desired point; otherwise one reflects A in the line getting A' and determines C as the intersection of the line with the segment A'B.

Hence: a light ray reflected in the line with angle of incidence and angle of reflection equal takes the shortest possible path from A to B via C. Or: if one wants to go from A to B on the shortest possible route and on the way fetch a pail of water from a (straight) river, then one must solve the same problem (COURANT-ROBBINS [16]).

The origins of the classical isoperimetric problem

3) among all plane closed curves of a given constant length (resp. "figures" of a given perimeter, isoperimetric figures) find the one which encloses the greatest possible area

go back at least to ZENODORES (between 200 BC and 90 AD). From him we have the following assertions (try to prove these with elementary or analytical geometry):

i) Given two regular polygons of the same perimeter the one with the larger number of sides has the greater enclosed area.

ii) If a circle has the same perimeter as a regular polygon, then the circle encloses the greater area.

iii) Among all isoperimetric triangles with the same base the isosceles triangle has the greatest area.

iv) Among all isoperimetric polygons with a given number of sides the regular one has the greatest area.

Thus the circle has a greater area than any isoperimetric polygon. STEINER (1836) showed: if the isoperimetric problem 3) has a solution, then it is necessarily a circle. The existence of a solution was only proved much later. While the existence is evident for the problems 1) and 2), it is by no means evident for the isoperimetric problem 3) (not to mention the spatial analog).

After appropriate generalization the following geometric problem is also of considerable interest to economists:

4) Suppose given three points in the plane. Find a point such that the sum of the distances from this point to the three given points is a minimum.

This problem appears to have been formulated for the first time by FERMAT in 1629. The names TORRICELLI and STEINER are also associated with the problem. The generalization to m points in $\mathbb{R}^n$ is often called the FERMAT-WEBER problem:

Suppose given m pairwise distinct points $a^1,\dots,a^m \in \mathbb{R}^n$ and positive weights $w_1,\dots,w_m$. Find a point $\bar{x} \in \mathbb{R}^n$ which minimizes the function $f : \mathbb{R}^n \to \mathbb{R}$ defined by

$$f(x) := \sum_{i=1}^{m} w_i |x-a^i|.$$

Here $|\ |$ denotes the euclidean norm, i.e. $|y| := \left(\sum_{i=1}^{n} y_i^2\right)^{1/2}$.

The economic interpretation is roughly the following: a chain of department stores wants to build a warehouse at a site such that the transportation costs to and from the warehouse are minimal. If $a^1,\dots,a^k \in \mathbb{R}^2$ are the locations of the stores and $a^{k+1},\dots,a^m \in \mathbb{R}^2$ the locations of the suppliers, then

$$\sum_{i=1}^{m} |x-a^i|$$

is the sum of the distance of a warehouse with location x to all the stores and all the suppliers. Taking account of differing transportation costs and differing quantities of goods leads to assigning the distances differing weights, i.e. in the simplest case to a cost function

$$f(x) := \sum_{i=1}^{m} w_i |x-a^i| .$$

We shall now demonstrate that the FERMAT-WEBER problem has exactly one solution if the given points $a^1,\dots,a^m$ are not collinear, i.e. do not all lie on one line.

α) Existence: Let $x^o \in \mathbb{R}^n$ be arbitrary. Then

$$W_o := \{x \in \mathbb{R}^n \;:\; f(x) \le f(x^o)\}$$

is compact. Thus the problem of minimizing the continuous function f on W_o has a solution $\bar{x}$, and since no solution can lie outside of W_o we have $f(\bar{x}) = \min\{f(x) : x \in \mathbb{R}^n\}$.

β) Uniqueness: Let x^1, x^2 be solutions, i.e.

$$f(x^1) = f(x^2) = \min\{f(x) \;:\; x \in \mathbb{R}^n\}$$

and $x^1 \neq x^2$. Then we have

$$\begin{aligned}
\min\{f(x) : x \in \mathbb{R}^n\} \le f(\tfrac{1}{2}(x^1+x^2)) &= \sum_{i=1}^{m} w_i |\tfrac{1}{2}(x^1-a^i)+\tfrac{1}{2}(x^2-a^i)| \\
&\le \frac{1}{2} \sum_{i=1}^{m} w_i (|x^1-a^i|+|x^2-a^i|) \\
&= \frac{1}{2} (f(x^1)+f(x^2)) = \min\{f(x) : x \in \mathbb{R}^n\}.
\end{aligned}$$

Hence $|(x^1-a^i)+(x^2-a^i)| = |x^1-a^i| + |x^2-a^i|$ and therefore

$$x^1 - a^i = \lambda_i(x^2-a^i) \text{ with } \lambda_i > 0 \ (i=1,\dots,m).$$

Since $x^1 \neq x^2$ we have $\lambda_i \neq 1$ and so

$$a^i - a^j = \frac{\lambda_i-\lambda_j}{(1-\lambda_i)(1-\lambda_j)} (x^1-x^2) \text{ for } i,j = 1,\dots,m.$$

Thus contrary to our assumption $a^1,\dots,a^m$ all lie on one line. It follows that $x^1 = x^2$; our solution is unique.

We now wish to give necessary and sufficient optimality conditions for a solution $\bar{x}$. We distinguish two cases:

i) $\bar{x} \in \{a^1,\dots,a^m\}$, say $\bar{x} = a^j$. For every $h \in \mathbb{R}^n$ with $|h| = 1$ we have:

$$0 \le \lim_{t\to 0+} \frac{1}{t} (f(a^j+th)-f(a^j))$$

$$= R_j^T h + w_j \text{ with } R_j = \sum_{\substack{i=1\\ i\neq j}}^{m} w_i \frac{a^j-a^i}{|a^j-a^i|}$$

It follows that

$$0 \le -|R_j| + w_j \quad \text{resp.} \quad |R_j| \le w_j.$$

We record for the moment: if f assumes its minimum at $\bar{x} = a^j$, then $|R_j| \le w_j$ must hold.

But this condition is also sufficient to quarentee a minimum at a^j. For if $|R_j| \le w_j$ and $x \in \mathbb{R}^n$ is arbitrary, then

$$f(x) - f(a^j) = \sum_{\substack{i=1\\ i\neq j}}^{m} w_i(|x-a^i|-|a^j-a^i|) + w_j|x-a^j|$$

$$\ge \sum_{\substack{i=1\\ i\neq j}}^{m} w_i\left(\frac{(a^j-a^i)^T(x-a^i)}{|a^j-a^i|} - |a^j-a^i|\right) + w_j|x-a^j|$$

$$= \sum_{\substack{i=1\\ i\neq j}}^{m} w_i \frac{(a^j-a^i)^T(x-a^j)}{|a^j-a^i|} + w_j|x-a^j|$$

$$= R_j^T(x-a^j) + w_j|x-a^j| \geq (-|R_j|+w_j)|x-a^j| \geq 0.$$

Now let $m = 3$ and $w_1 = w_2 = w_3 = 1$. We wish to show that in this case $|R_j| \leq 1$ is equivalent to the statement that the triangle with vertices a^1, a^2, a^3 has an angle greater than or equal to $120°$ at the vertex a^j.

Suppose for example $j = 2$. Then

$$|R_2|^2 = \left|\frac{a^2-a^1}{|a^2-a^1|} + \frac{a^2-a^3}{|a^2-a^3|}\right|^2 = 2 + 2\,\frac{(a^1-a^2)^T(a^3-a^2)}{|a^1-a^2|\,|a^3-a^2|}$$

Thus $|R_2| \leq 1$ precisely when $\cos \sphericalangle\,(a^1-a^2, a^3-a^2) \leq -\frac{1}{2}$ resp. when in $\Delta a^1a^2a^3$ the angle at a^2 is greater than or equal $120°$. Thus if $\Delta a^1a^2a^3$ has an angle greater than or equal to $120°$, then the corresponding vertex is the solution of the FERMAT problem; otherwise the solution cannot lie at a vertex.

ii) $\bar{x} \notin \{a^1,\dots,a^m\}$. Then $f(x) = \sum_{i=1}^{m} w_i|x-a^i|$ is continuously differentiable at $\bar{x}$ and we necessarily have

$$\nabla f(\bar{x}) = \sum_{i=1}^{m} w_i \frac{\bar{x}-a^i}{|\bar{x}-a^i|} = 0.$$

With

$$\lambda_i := \frac{w_i}{|\bar{x}-a^i|} \Big/ \sum_{j=1}^{m} \frac{w_j}{|\bar{x}-a^j|}$$

we then have

$$\bar{x} = \sum_{i=1}^{m} \lambda_i a^i.$$

Because

$$\lambda_i > 0 \ (i=1,\dots,m) \text{ and } \sum_{i=1}^{m} \lambda_i = 1$$

it follows that $\bar{x}$ is a convex combination of the a^i resp. lies in the convex hull of the a^i and thus for $m = 3$ lies in the triangle $\Delta a^1a^2a^3$. Furthermore for $j = 1,\dots,m$:

$$0 = \nabla f(\bar{x})^T \frac{\bar{x}-a^j}{|\bar{x}-a^j|} = \sum_{i=1}^{m} w_i \frac{(\bar{x}-a^i)^T(\bar{x}-a^j)}{|\bar{x}-a^i|\;|\bar{x}-a^j|}$$

$$= \sum_{i=1}^{m} w_i \cos \alpha_{ij},$$

where $\alpha_{ij} := \sphericalangle\,(a^i-\bar{x}, a^j-\bar{x})$.

For m = 3 one can easily determine $\cos \alpha_{12}$, $\cos \alpha_{13}$ and $\cos \alpha_{23}$ from the equations

$$\begin{aligned} w_2 \cos \alpha_{12} + w_3 \cos \alpha_{13} \qquad\qquad &= - w_1 \\ w_1 \cos \alpha_{12} \qquad\qquad + w_3 \cos \alpha_{23} &= - w_2 \\ w_1 \cos \alpha_{13} + w_2 \cos \alpha_{23} &= - w_3 \end{aligned}$$

and for $w_1 = w_2 = w_3 = 1$ for example one obtains

$$\cos \alpha_{12} = \cos \alpha_{13} = \cos \alpha_{23} = -\frac{1}{2}.$$

In this case $\bar{x}$ is necessarily a point in $\Delta a^1a^2a^3$ in which the three sides of the triangle subtend an angle of 120°. Such a point is called a FERMAT- or TORRICELLI-point.

We now have a complete solution of the FERMAT problem: if an angle of the triangle is greater than or equal 120° then the corresponding vertex is the solution; otherwise it is the FERMAT-point.

We conclude this section by considering a further classical problem, which was posed and solved by J.F. de TUSCHIS a FAGNANO in 1775 and which is often also called the SCHWARZ triangle problem in honor of H.A. SCHWARZ.

5) Find a point on each of the sides of a given acute triangle such that the triangle determined by these points has minimal perimeter.

The vertices of the desired triangle are the bases of the altitudes of the given triangle. There are many interesting proofs of this fact; we shall give one which resembles the proofs of LHUILIER (1809) and LINDELÖF (1870) but refer the reader to a very elegant elementary geometric proof by L. FEJER (see e.g. COXETER [18, p. 20] and RADEMACHER-TOEPLITZ

[66]; further references in COURT [17]).

Suppose the vertices A,B,C of the given triangle have coordinates a,b,c. For the coordinates u,v,w of the vertices U,V,W we seek let us try taking

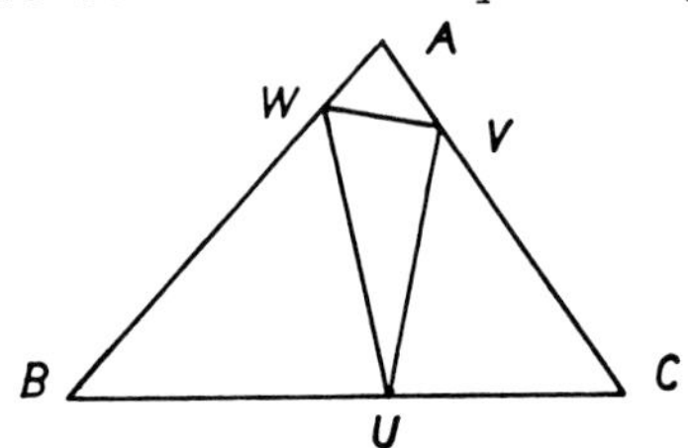

$$u = \lambda b + (1-\lambda)c$$
$$v = \mu c + (1-\mu)a$$
$$w = \nu a + (1-\nu)b.$$

The problem of minimizing

$$\begin{aligned} f(\lambda,\mu,\nu) &:= |u-v| + |v-w| + |w-u| \\ &= |\lambda(b-c)+(1-\mu)(c-a)| + |\mu(c-a)+(1-\nu)(a-b)| \\ &\quad + |\nu(a-b)+(1-\lambda)(b-c)| \end{aligned}$$

under the side condition $(\lambda,\mu,\nu) \in M := [0,1] \times [0,1] \times [0,1]$ has a solution (λ,μ,ν) since M is compact and f continuous. Let us first suppose that $(\lambda,\mu,\nu) \in \text{int } M = (0,1) \times (0,1) \times (0,1)$. Then the gradient of f vanishes at (λ,μ,ν), i.e.

i) $\frac{\partial f}{\partial \lambda}(\lambda,\mu,\nu) = \left\{ \frac{u-v}{|u-v|} - \frac{w-u}{|w-u|} \right\}^T (b-c) = 0$

ii) $\frac{\partial f}{\partial \mu}(\lambda,\mu,\nu) = \left\{ \frac{v-w}{|v-w|} - \frac{u-v}{|u-v|} \right\}^T (c-a) = 0$

iii) $\frac{\partial f}{\partial \nu}(\lambda,\mu,\nu) = \left\{ \frac{w-u}{|w-u|} - \frac{v-w}{|v-w|} \right\}^T (a-b) = 0$

From these equations we conclude successively

i) $\sphericalangle$ VUC = $\sphericalangle$ WUB
ii) $\sphericalangle$ WVA = $\sphericalangle$ UVC
iii) $\sphericalangle$ UWB = $\sphericalangle$ VWA

and hence

$\sphericalangle$ VUC = $\sphericalangle$ WUB = $\sphericalangle$ CAB =: α
$\sphericalangle$ WVA = $\sphericalangle$ UVC = $\sphericalangle$ ABC =: β
$\sphericalangle$ UWB = $\sphericalangle$ VWA = $\sphericalangle$ BCA =: γ

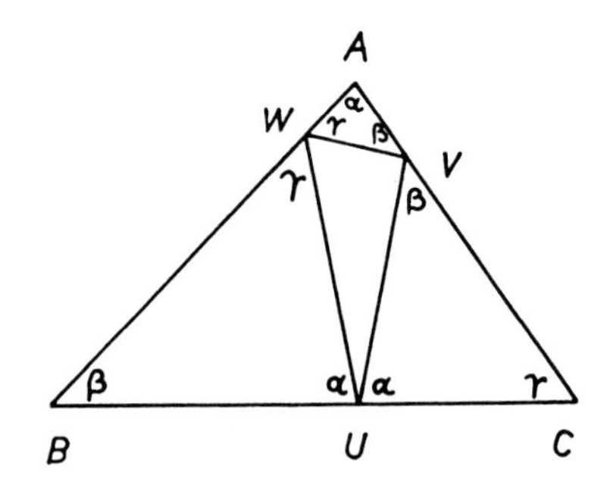

Now we conclude that ΔUVW is necessarily an orthic triangle.

Applying the law of sines gives:

$$\frac{BC}{CA} = \frac{\sin\alpha}{\sin\beta} = \frac{CV}{CU} = \frac{CA-AV}{CU}$$

$$\frac{CA}{AB} = \frac{\sin\beta}{\sin\gamma} = \frac{AW}{AV} = \frac{AB-BW}{AV}$$

$$\frac{AB}{BC} = \frac{\sin\gamma}{\sin\alpha} = \frac{BU}{BW} = \frac{BC-CU}{BW}$$

For CU, AV, BW one thus gets a system of linear equations

$$\begin{aligned} BC\cdot CU + CA\cdot AV \qquad\qquad &= CA^2 \\ CA\cdot AV + AB\cdot BW &= AB^2 \\ BC\cdot CU \qquad\qquad + AB\cdot BW &= BC^2\ , \end{aligned}$$

and thus

$$CU = \frac{BC^2+CA^2-AB^2}{2BC} = CA\cdot\cos\gamma$$

$$AV = \frac{CA^2+AB^2-BC^2}{2CA} = AB\cdot\cos\alpha$$

$$BW = \frac{AB^2+BC^2-CA^2}{2AB} = BC\cdot\cos\beta$$

from which one can read off that AU is the altitude in ΔACU, BV the altitude in ΔBAV and CW the altitude in ΔCBW, i.e. ΔUVW is orthic. As circumference of UVW we get

$$\begin{aligned} \mathrm{per}(UVW) &= UV + VW + WU \\ &= CU\cdot\frac{AB}{CA} + AV\cdot\frac{BC}{AB} + BW\cdot\frac{CA}{BC} \\ &\qquad \text{(law of sines)} \\ &= AB\cos\gamma + BC\cdot\cos\alpha + CA\cdot\cos\beta \end{aligned}$$

$$= 2r(\sin\gamma\cos\gamma + \sin\alpha\cdot\cos\alpha + \sin\beta\cos\beta)$$

$$= 4r\ \sin\alpha\sin\beta\sin\gamma$$

where r is the radius of the circumscribing circle of ABC. From this one can also see that the optimal triangle is necessarily nondegenerate, i.e. that none of the desired points U,V,W coincides with a vertex of the given triangle. For the "circumference" of the smallest degenerate triangle is twice the smallest altitude, say

$$2CA\ \sin\alpha = 4r\ \sin\alpha\sin\beta > \mathrm{per}(UVW) = 4r\ \sin\alpha\sin\beta\sin\gamma$$

since we assumed that the given triangle was acute.

1.2 Calculus of variations

We cannot attempt here to present the history of the calculus of variations. A reader who wants to know more about that topic should turn to GOLDSTINE [29] or the introduction to BLANCHARD-BRUENING [5]. We shall merely present several historically important problems: we shall come back to their solution later, when the necessary tools are available.

1) JOHANN BERNOULLI posed in 1696 as a task for the "most ingenious mathematicians of the world" the problem of the brachystochrone: find the curve joining two given points P_o and P_1 not lying vertically one above the other with the property that a point mass sliding without friction along this curve from P_o to P_1 only under the influence of gravity will reach P_1 in the shortest possible time.

If $P_o = (t_o, x_o)$, $P_1 = (t_1, x_1)$ with $t_o < t_1$, $x_1 < x_o$, then the set of admissible curves, i.e. "smooth curves", which connect P_o and P_1 is given by

$$M := \{x \in C[t_o,t_1] \cap C^1(t_o,t_1] : x(t_o) = x_o, x(t_1) = x_1\}$$

(one does not insist on differentiability at t_o since one could imagine that the desired curve x is not differentiable at t_o). The time of traversal for a curve $x \in M$ is given by

$$T(x) = \frac{1}{\sqrt{2g}} \int_{t_0}^{t_1} \sqrt{\frac{1+x'^2(t)}{x_0-x(t)}}\, dt$$

Already GALILEI had found by experiment, that if one lets two identical balls start simultaneously, the one along an arc of a circle, the other along the corresponding secant, then the one on the circle reaches the lower end first.

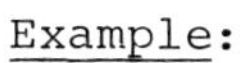

Example:

$$\text{Let } x_1(t) = -t \qquad \text{(secant)}$$

$$x_2(t) = -\sqrt{1-(1-t)^2} \qquad \text{(arc of circle).}$$

Then

$$T(x_1) = \frac{2}{\sqrt{g}}$$

$$T(x_2) = \frac{1}{\sqrt{2g}} \int_0^1 \frac{dt}{(1-(1-t)^2)^{3/4}}$$

$$= \frac{1}{\sqrt{2g}} \int_0^1 \frac{ds}{\sqrt{s(1-s^2)}} \qquad (s^2 = 1-(1-t)^2)$$

$$= \frac{1}{\sqrt{g}} \int_0^{\pi/2} \frac{d\varphi}{\sqrt{1-\frac{1}{2}\sin^2\varphi}} \qquad \left(s = \frac{\frac{1}{2}\sin^2\varphi}{1-\frac{1}{2}\sin^2\varphi}\right)$$

$$\approx \frac{1.854009}{\sqrt{g}}$$

Thus in this example the time of traversal along the arc of the circle is indeed shorter than that along the secant.

As solution of the brachystochrone problem one obtains cycloids, which can be given in parameter form as follows:

$$\left.\begin{aligned} t &= t_0 + \frac{A}{2}(\varphi - \sin\varphi) \\ x &= x_0 - \frac{A}{2}(1 - \cos\varphi) \end{aligned}\right\} \quad 0 \le \varphi \le \varphi_1$$

The still undetermined parameters A, φ_1 are determined by the end condition

$$t_1 = t_o + \frac{A}{2} (\varphi_1 - \sin \varphi_1)$$

$$x_1 = x_o - \frac{A}{2} (1 - \cos \varphi_1)$$

For the example given above this means that one has to solve the system of equations

$$1 = \frac{A}{2} (\varphi_1 - \sin \varphi_1)$$

$$-1 = -\frac{A}{2} (1 - \cos \varphi_1)$$

From these one obtains $A \approx 1.145834$, $\varphi_1 \approx 2.412011$ and as traversal time

$$T_{min} = \frac{1}{\sqrt{g}} \sqrt{\frac{A}{2}}\, \varphi_1$$

$$\approx \frac{1.825682}{\sqrt{g}}$$

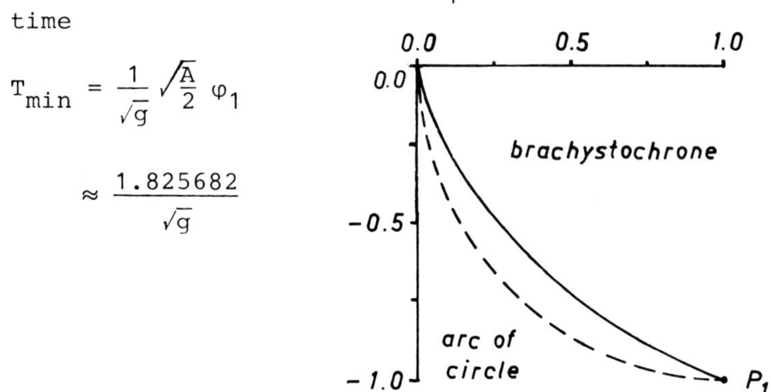

2) The problem of geodetic lines consists in finding the shortest curve lying entirely on a surface S in $\mathbb{R}^3$ and connecting two given points P_o and P_1 in S.

If $\varphi(x,y,z) = 0$ is the equation of the surface S and if we consider the curves to be given in parameter form by $(x(t),y(t),z(t))$ for $t_o \le t \le t_1$, then the problem consists in minimizing

$$I(C) := \int_{t_o}^{t_1} \sqrt{\dot{x}^2(t)+\dot{y}^2(t)+\dot{z}^2(t)}\,dt$$

with the side conditions

$$\varphi(x(t),y(t),z(t)) = 0 \quad \text{for} \quad t \in [t_o,t_1]$$

$$P_o = (x(t_o),y(t_o),z(t_o)), \quad P_1 = (x(t_1),y(t_1),z(t_1))$$

3) According to legend the princess Dido of Tyros fled to north Africa after her brother had murdered her husband; there she asked King Hierbas for as much land as she could surround with the hide of a cow. When he granted her her wish, she cut the hide into thin strips and stretched her thong around a gigantic piece of land. Obviously Dido had to solve the isoperimetric problem: find the region with maximal area and given fixed circumference.

We consider a simpler problem: given are numbers a,l with $0 < a < l$. Find the curve of length 2l lying in the upper half plane, connecting the points $A = (-a,0)$ and $B = (a,0)$ and together with the line segment AB enclosing the largest possible area. Thus one has to solve the following problem: maximize

$$I(y) = \int_{-a}^{a} y(x)\,dx$$

on the set

$$M := \{y \in C^1[-a,a] \ : \ y(-a) = 0 = y(a), \int_{-a}^{a} \sqrt{1+y'^2(x)}\,dx = 2l\}.$$

To solve the problem one must distinguish two cases.

i) $a < l < \frac{1}{2}\pi a$: Then one obtains as solution an arc of a circle:

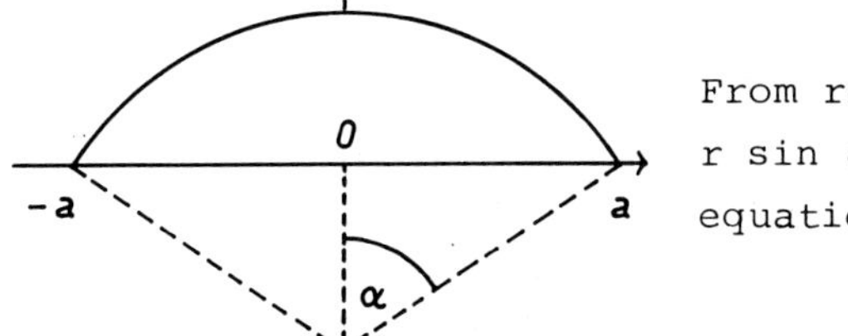

From $r\alpha = l$ and $r \sin \alpha = a$ one gets the equation $\frac{\sin \alpha}{\alpha} = \frac{a}{l}$ for α

ii) $l > \frac{1}{2}\pi a$: In this case the problem above has no solution. We have

$$\sup_{y \in M} I(y) = 2al - \frac{1}{2}\pi a^2,$$

but the supremum is not achieved by any $\bar{y} \in M$. The "bounding curve" is not of the form $y = y(x)$:

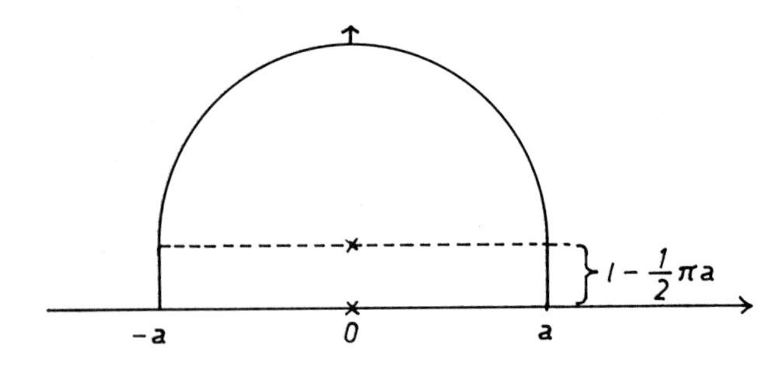

1.3 Approximation problems

Approximation problems are roughly speaking special optimization problems in which the objective function is given by a norm. More precisely:

Let $(X, \| \ \|)$ be a linear normed space, $z \in X$ and $V \subset X$. Then the optimization problem

$$\text{minimize } f(x) := \| x-z \| \quad \text{on } V$$

is called an approximation problem. A solution $\bar{x}$, i.e. an element $\bar{x} \in V$ with $\| \bar{x}-z \| \le \| x-z \|$ for all $x \in V$ is called the best approximation for z with respect to V.

1) The method of least squares -already applied by GAUSS around 1800- has to do with the following: let $z = (z_1,\dots,z_m)^T \in \mathbb{R}^m$ be a known vector, for example m measurements obtained in m successive experiments. One assumes that the z_i depend functionally, e.g. linearly, on a certain parameter vector $y = (y_1,\dots,y_n)^T \in \mathbb{R}^n$, $n < m$, which is to be determined:

$$z_i = g_i(y_1,\dots,y_n) \quad (i=1,\dots,m) \quad \text{resp. } z = g(y).$$

Since the z_i are subject to errors of measurement resp. the system of equations $z = g(y)$ is overdetermined since $n < m$, one seeks with the method of least squares a $\bar{y} \in \mathbb{R}^n$ such that the defect $z - g(y)$ is minimal with respect to the euclidean norm. Thus one has to solve an approximation problem with $(X, \| \ \|) = (\mathbb{R}^m, | \ |)$ and $V = g(\mathbb{R}^n)$.

If g is linear, i.e. $g(y) = Ay$ with $A \in \mathbb{R}^{m\times n}$, then one must minimize $f(y) := |z-Ay|$ on $\mathbb{R}^n$.

2) A CHEBYSHEV approximation problem is one in which $X = C(B)$ is the linear space of continuous real-valued functions on the compact set $B \subset \mathbb{R}^N$ and the norm $\| \quad \|$ is given by the maximum norm:

$$\| x \|_\infty := \max_{t \in B} |x(t)| .$$

Here it is a question of approximating a given function $z \in C(B)$ with "easily" computable functions $x \in V$ (e.g. polynomials or rational functions of a given degree), so that the maximal defect in norm

$$\max_{t \in B} |x(t)-z(t)|$$

is minimal. These problems became particularly important for applications, when it became possible and necessary to program subroutines for the calculation of elementary functions (e.g. roots, trigonometric functions) on computers and pocket calculators.

As an example we consider the calculation of the square root $z(t) = \sqrt{t}$ for $t > 0$. The first step is a reduction with the object of restricting the approximation to a finite interval. Let $t = m\beta^e$ be a positive floating point number with mantissa m $(\frac{1}{\beta} \le m < 1)$ to base β (e.g. $\beta = 2$, 10 or 16) and exponent e. Then

$$\sqrt{t} = \begin{cases} \sqrt{m}\ \beta^{e/2} & \text{(e even)} \\ \frac{1}{\sqrt{\beta}}\ \sqrt{m}\ \beta^{(e+1)/2} & \text{(e odd)} \end{cases}$$

Thus we have reduced the calculation of $\sqrt{t}$ to the calculation of $\sqrt{m}$ for $\frac{1}{\beta} \le m < 1$, although for odd exponent e one still has to devide by $\sqrt{\beta}$. One can avoid this annoyance if one proceeds from the presentation $t = M\beta^{2E}$ with $1/\beta^2 \le M < 1$. Because $\sqrt{t} = \sqrt{M}\beta^E$ one has to calculate $\sqrt{M}$ for $1/\beta^2 \le M < 1$. In practice (c.f. FIKE [23], MOURSUND [61]) approximations of $\sqrt{t}$ have been considered chiefly on the intervals $[\frac{1}{16},1]$, $[\frac{1}{4},1]$,

$[\frac{1}{2},1]$, i.e. on an interval $I = [a,1]$ with $a \in (0,1)$.

In a second step we determine an easily computable approximation $x_o(t)$ for $\sqrt{t}$ on the interval $I = [a,1]$. In the simplest case, and we shall consider only this one, $x_o(t) = A + Bt$ is a linear function. The coefficients A,B are chosen such that the maximal relative error with respect to the absolute value

$$f(A,B) := \max_{t\in I} \left|\frac{x_o(t)-\sqrt{t}}{\sqrt{t}}\right| = \max_{t\in I} \left|\frac{A+Bt-\sqrt{t}}{\sqrt{t}}\right|$$

is minimal. This is the CHEBYSHEV approximation problem to approximate the constant function $z(t) = 1$ by elements of $V = \mathrm{span}(v_1,v_2)$ with $v_1(t) = t^{-1/2}$, $v_2(t) = t^{1/2}$. The solution is

$$\bar{A} = \frac{2a^{1/2}}{(1+a^{1/4})^2}, \quad \bar{B} = \frac{2}{(1+a^{1/4})^2}, \quad f(\bar{A},\bar{B}) = \left(\frac{1-a^{1/4}}{1+a^{1/4}}\right)^2$$

The precision of $x_o(t) = \bar{A} + \bar{B}t$ as approximation to $\sqrt{t}$ is naturally not good enough yet, so in a third step we apply NEWTON iteration, i.e. set

$$x_{m+1}(t) = \frac{1}{2}\left(x_m(t) + \frac{t}{x_m(t)}\right) \qquad (m=0,1,2,\ldots)$$

If one defines the relative error of the mth iterate by

$$\delta_m(t) := \frac{x_m(t)-\sqrt{t}}{\sqrt{t}}$$

then one has

$$\delta_{m+1}(t) = \frac{\delta_m^2(t)}{2(1+\delta_m(t))} \geq 0.$$

E.g. $\| \delta_2 \|_\infty \leq 0.4 \cdot 10^{-10}$ for $a = \frac{1}{2}$.

A more subtile method for determining the starting approximation $x_o(t) = \bar{A} + \bar{B}t$ for a NEWTON iteration has been proposed by MOURSUND [61]. $\bar{A},\bar{B}$ are chosen so that the relative error of the mth NEWTON iterate $x_m(t)$ is minimal. Thus with $x_o(t) = A + Bt$,

$$x_{m+1}(t) = \frac{1}{2}\left(x_m(t) + \frac{t}{x_m(t)}\right)$$

and relative error

$$\delta_m(t) = \frac{x_m(t)-\sqrt{t}}{\sqrt{t}}$$

we have to minimize

$$f_m(A,B) := \max_{t\in I} |\delta_m(t)|.$$

If one defines the logarithmic relative error by

$$\hat{\delta}_m(t) := \ln(1+\delta_m(t)) = \ln \frac{x_m(t)}{\sqrt{t}}$$

and thus

$$\hat{f}_m(A,B) := \max_{t\in I} |\hat{\delta}_m(t)|,$$

then we get the following remarkable result (c.f. MOURSUND [61], KING-PHILLIPS [41]):

if $\hat{f}_o(\bar{A},\bar{B}) \leq \hat{f}_o(A,B)$ for all $A,B \in \mathbb{R}$, then for

$m = 1,2,\ldots$: $\hat{f}_m(\bar{A},\bar{B}) \leq \hat{f}_m(A,B)$ and

$f_m(\bar{A},\bar{B}) \leq f_m(A,B)$ for all $A,B \in \mathbb{R}$.

It thus suffices to minimize the logarithmic relative error of the linear starting approximation. A solution of this problem simultaneously minimizes the maximal relative error of all successive NEWTON iterates.

In fact:

i) $$\delta_{m+1}(t) = \frac{1}{2}\left\{(1+\delta_m(t)) + \frac{1}{1+\delta_m(t)}\right\} - 1 \geq 0$$

for $m = 0,1,2,\ldots$. From this one gets the recursion formula

$$\hat{\delta}_{m+1}(t) = \ln(1+\delta_{m+1}(t)) = \ln\cosh \hat{\delta}_m(t) \geq 0$$

for m = 0,1,2,... . Because $g(s) := \ln\cosh s$ is an even function and for positive s monotone increasing we have

$$\hat{f}_{m+1}(A,B) = \ln\cosh \hat{f}_m(A,B)$$

From $\hat{f}_o(\bar{A},\bar{B}) \le \hat{f}_o(A,B)$ for all $A,B \in \mathbb{R}$ it thus also follows that $\hat{f}_m(\bar{A},\bar{B}) \le \hat{f}_m(A,B)$ for all $A,B \in \mathbb{R}$ and m = 0,1,2,... .

ii) Suppose there were an $m \in \mathbb{N}$ and $A',B' \in \mathbb{R}$ with $f_m(A',B') < f_m(A,B)$. Let $\delta'_m(t)$ resp. $\delta_m(t)$ be the relative errors of the mth NEWTON iterates belonging to A',B' resp. $\bar{A},\bar{B}$. Because $\delta'_m(t),\delta_m(t) \ge 0$ we have

$$f_m(A',B') = \max_{t\in I} \delta'_m(t) < \max_{t\in I} \delta_m(t) = f_m(\bar{A},\bar{B})$$

Since $h(s) := \ln(1+s)$ is monotone increasing we have

$$\hat{f}_m(A',B') = \max_{t\in I} \ln(1+\delta'_m(t)) < \max_{t\in I} \ln(1+\delta_m(t)) = \hat{f}_m(\bar{A},\bar{B})$$

which is a contradiction to i).

It thus suffices to solve the problem

$$\text{minimize } \hat{f}_o(A,B) = \max_{t\in I} \left|\ln\left(\frac{A}{\sqrt{t}} + B\sqrt{t}\right)\right|.$$

The solution is

$$\bar{A} = \frac{a^{1/2}}{2^{1/2}a^{1/8}(1+a^{1/2})^{1/2}}, \quad \bar{B} = \frac{1}{2^{1/2}a^{1/8}(1+a^{1/2})^{1/2}}$$

and we have

$$\hat{f}_o(\bar{A},\bar{B}) = \frac{1}{2}\ln\left(\frac{1+a^{1/2}}{2a^{1/4}}\right), \quad \hat{f}_{m+1}(\bar{A},\bar{B}) = \ln\cosh \hat{f}_m(\bar{A},\bar{B})$$

$$f_1(\bar{A},\bar{B}) = \frac{1}{2}\,\frac{((1+a^{1/2})^{1/2}-2^{1/2}a^{1/8})^2}{2^{1/2}a^{1/8}(1+a^{1/2})^{1/2}}, \quad f_{m+1}(\bar{A},\bar{B}) = \frac{f_m^2(\bar{A},\bar{B})}{2(1+f_m(\bar{A},\bar{B}))}$$

It is somewhat disappointing that what we gain by this method compared to the naive procedure described earlier is not very much. For $a = \frac{1}{2}$ and $m = 2$ for example the maximal relative error is about 1.4 % smaller.

1.4 Linear programming

These days it is no longer necessary to emphasize the great importance of linear programming for applications. Probably no other branch of mathematics has had a similarly great influence in so many areas, particularly in industry and commerce. We shall not go into the historical development of the subject here, a development connected with such names as KANTOROWICZ (1939), HITCHCOCK (1941) and DANTZIG (1947). We refer the reader to the fundamental book by DANTZIG [19].

By a (finite dimensional) linear optimization problem we mean the following:

minimize a real-valued linear function f on a subset M of $\mathbb{R}^n$ which is given by finitely many linear equations and inequalities.

Since some of the constraints are often given in the form of sign restrictions for some of the variables, it is reasonable to treat these variables separately. A general linear programming problem is thus given as follows:

$$\text{minimize } f(x) = c_1^T x_1 + c_2^T x_2 \text{ on the set}$$

$$M := \{x = (x_1, x_2) \in \mathbb{R}^{n_1} \times \mathbb{R}^{n_2} : \begin{array}{l} A_{11}x_1 + A_{12}x_2 = b_1 \\ A_{21}x_1 + A_{22}x_2 \geq b_2 \end{array}, \ x_1 \geq 0\}.$$

Here $c_j \in \mathbb{R}^{n_j}$, $b_i \in \mathbb{R}^{m_i}$ and $A_{ij} \in \mathbb{R}^{m_i \times n_j}$ for $i,j = 1,2$. $n = n_1 + n_2$ is the number of variables, $m = m_1 + m_2$ the number of equations and inequalities.

In the numerous text books about linear programming one finds many examples ranging from simple two dimensional ones (the number of variables n is two), which can be solved graphically, to rather complicated ones, where setting up the mathematical model already costs a fair amount of work. Here we wish to give several fundamental examples.

1) The diet problem formulated by STIGLER in 1939 is the following: one wishes to set up a diet chosen from n foods (e.g. wheat flour, condensed milk, cheddar cheese, beef liver, green beans, cabbage, etc. STIGLER has n = 77), containing m nutrients (e.g. protein, calcium, iron, vitamines A, B_1, B_2, C etc.), which on the one hand must supply the minimum requirements of the basic nutrients and on the other hand must be as cheap as possible.

A diet plan consists of a vector $x = (x_1,\dots,x_n)^T$, where x_j gives the number of units of the jth food.

If a_{ij} is the number of units of the ith nutrient in the jth food and b_i the minimum requirement of the ith nutrient, then a diet plan $x = (x_1,\dots,x_n)^T$ is feasible if

$$\sum_{j=1}^{n} a_{ij}x_j \geq b_i \quad (i=1,\dots,m), \quad x_j \geq 0 (j=1,\dots,n)$$

If furthermore c_j is the price of one unit of the jth food, then the cost of a diet plan x is given by

$$f(x) = \sum_{j=1}^{n} c_j x_j .$$

If we introduce the matrix $A = (a_{ij}) \in \mathbb{R}^{m\times n}$ and the vectors $b = (b_i) \in \mathbb{R}^m$ and $c = (c_j) \in \mathbb{R}^n$, then the diet problem can be formulated:

$$\text{minimize } f(x) = c^T x \text{ on } M := \{x \in \mathbb{R}^n : Ax \geq b, x \geq 0\}.$$

Here we have used the following suggestive notation: for $x,z \in \mathbb{R}^n$ $x \geq z$ shall mean that $x_j \geq z_j$ for $j = 1,\dots,n$. 0 means the zero element of the corresponding linear space, in this case $\mathbb{R}^n$.

2) The production planning problem differs as mathematical problem only slightly from the diet problem: a factory (e.g. a dairy) produces n products (e.g. butter, buttermilk, milk powder, gouda cheese, edam cheese and whey) and uses m resources (e.g. raw milk, machines, storage rooms). The profit is to be maximized under the side condition that the availability

of the resources is not unlimited.

A production plan consists of a vector $x = (x_1,\dots,x_n)^T$. This vector means that x_j units of the jth product are to be produced. One needs a_{ij} units of the ith resource in order to produce one unit of the jth product. Furthermore a maximum of b_i units of the ith resource are available. Thus a production plan $x = (x_1,\dots,x_n)^T$ is feasible if

$$\sum_{j=1}^{n} a_{ij}x_j \le b_i \ (i=1,\dots,m), \quad x_j \ge 0 \ (j=1,\dots,n).$$

If in addition the net profit by the production of one unit of product j is p_j, then the total net profit by application of the production plan $x = (x_1,\dots,x_n)^T$ is given by

$$\sum_{j=1}^{n} p_j x_j .$$

With $A = (a_{ij}) \in \mathbb{R}^{m\times n}$, $b = (b_i) \in \mathbb{R}^m$ and $p = (p_j) \in \mathbb{R}^n$ we can formulate the production planning problem as follows:

$$\text{maximize } p^T x \text{ on } M := \{x \in \mathbb{R}^n : Ax \le b,\ x \ge 0\}.$$

3) The <u>transportation problem</u> (formulated by HITCHCOCK in 1941) is the following: m refineries $R_1,\dots,R_m$ of an oil company supply n storage depots $T_1,\dots,T_n$. The refinery R_i can produce at most r_i units of a particular product, the depot T_j needs at least t_j units. Transportation from R_i to T_j costs c_{ij} units of money per unit of the product. One seeks a transportation plan which with minimal costs fills the needs of the depots without exceeding the production limits of the refineries.

A transportation plan consists of an m×n matrix $x = (x_{ij})$, where x_{ij} means the amount to be transported from R_i to T_j. The plan is feasible if

$$\sum_{j=1}^{n} x_{ij} \le r_i \qquad (i=1,\dots,m)$$

(i.e. the amount transported from R_i is not greater than the production limit), and

$$\sum_{i=1}^{n} x_{ij} \geq t_j \qquad (j=1,\ldots,n)$$

(i.e. the amount arriving at T_j covers the needs) and

$$x_{ij} \geq 0 \ (i=1,\ldots,m, \ j=1,\ldots,n).$$

Since

$$\sum_{i=1}^{m} \sum_{j=1}^{n} c_{ij}x_{ij}$$

is the cost of the transportation plan, we have the problem:

$$\text{minimize } f(x) := \sum_{i=1}^{m} \sum_{j=1}^{n} c_{ij}x_{ij} \quad \text{on}$$

$$M := \{x = (x_{ij}) \in \mathbb{R}^{m\times n} : \sum_{j=1}^{n} x_{ij} \leq r_i, \ \sum_{i=1}^{m} x_{ij} \geq t_j \ ,$$

$$x_{ij} \geq 0 \text{ for } i = 1,\ldots,m, \ j = 1,\ldots,n\}.$$

One naturally assumes that r_i and t_j are nonnegative. Necessary and sufficient for the existence of a feasible plan, i.e. for $M \neq \emptyset$, is obviously the condition that

$$\sum_{i=1}^{m} r_i \geq \sum_{j=1}^{n} t_j$$

that is that the total production not be smaller than the total need.

4) In linear regression analysis using the method of least squares (c.f. 1.3) one has on over-determined system of linear equations $Ax = y$ with $A \in \mathbb{R}^{m\times n}$, $y \in \mathbb{R}^m$, $n < m$ and one wants a vector $\overline{x}$ whose defect with respect to the euclidean norm is minimal; that is one seeks a solution $\overline{x}$ of the problem: minimize $f(x) := |y-Ax|$. If instead the maximal deviation with respect to all coordinates is to be minimal, then one must minimize the defect with respect to the maximum norm

$$\| z \|_\infty := \max_{i=1,\ldots,m} |z_i|.$$

Thus one seeks an $\bar{x}$ for which $f(x) := \| y-Ax \|_\infty$ is minimal. This is really a linear programming problem in disguise, for it is equivalent to:

minimize δ under the side condition

$$\| y-Ax \|_\infty \leq \delta$$

respectively:

$$\text{minimize } \delta = \begin{pmatrix} 1 \\ 0 \end{pmatrix}^T \begin{pmatrix} \delta \\ x \end{pmatrix} \text{ on}$$

$$M := \left\{ \begin{pmatrix} \delta \\ x \end{pmatrix} \in \mathbb{R}^{1+n} : -\delta e \leq y - Ax \leq \delta e \right\},$$

where $e := (1,\ldots,1)^T \in \mathbb{R}^m$.

5) Finally we shall give yet another geometric problem, which one can formulate as a linear programming problem:

$$\text{Let } P := \{x \in \mathbb{R}^n : a^{iT}x \leq b_i \ (i=1,\ldots,m)\}$$

with $|a^i| = 1$ $(i=1,\ldots,m)$. Let P be bounded. Thus P is the intersection of m half spaces $\{x \in \mathbb{R}^n : a^{iT}x \leq b_i\}$ bounded by the hyperplanes $H := \{x \in \mathbb{R}^n : a^{iT}x = b_i\}$. We seek the ball with maximal radius that can be contained in P, the socalled insphere. If one denotes by $B[x;r]$ the closed ball with center x and radius r, i.e.

$$B[x;r] := \{y \in \mathbb{R}^n : |y-x| \leq r\},$$

then we have the problem

maximize r under the side condition $B[x;r] \subset P$.

But $B[x;r] \subset P \iff a^{iT}x - b_i + r \leq 0 \quad (i=1,\ldots,m), \ r \geq 0$

For if $B[x;r] \subset P$, then in particular $x + ra^i \in P$ for $i = 1,\ldots,m$ and thus $a^{iT}x + r \leq b_i$. On the other hand if $a^{iT}x - b_i + r \leq 0$ $(i=1,\ldots,m)$, $r \geq 0$ and $y \in B[x;r]$, then

$$a^{iT}y = a^{iT}x + a^{iT}(y-x)$$

$$\leq a^{iT}x + |y-x| \quad \text{(Cauchy-Schwarz and } |a^i| = 1)$$

$$\leq a^{iT}x + r \leq b_i \quad (i=1,\ldots,m).$$

Thus one obtains the insphere and in particular the inradius $\bar{r}$ as solution of the linear programming problem

maximize r subject to

$$a^{iT}x - b_i + r \leq 0 \quad (i=1,\ldots,m)$$

$$r \geq 0.$$

1.5 Optimal Control

Since about 1950 particularly motivated by problems of space travel one has considered the problem of controlling a system described for example by a system of ordinary differential equations in such a way that under given restrictions a objective function is minimized. Here we shall not attempt to formulate a general optimal control problem: instead we shall give two typical examples.

1) Suppose given the problem of bringing a vertically climbing rocket with adjustable thrust to a maximum altitude (c.f. e.g. TOLLE [73, p. 24], LEE-MARKUS [49, p. 456ff]).

Let $m = m(t)$ mass of the rocket at time t
$v = v(t)$ vertical velocity of the rocket at time t
$h = h(t)$ height of the rocket at time t.

One sets up the equations of motion as follows: in time Δt the rocket engines expel a mass Δm with a constant escape velocity w_A. At time $t + \Delta t$ the total system consists of the mass $m - \Delta m$ with velocity $v + \Delta v$ and the mass Δm with velocity $v - w_A$. If there are no external forces, then the law of conservation of momentum says:

$$mv = (m-\Delta m)(v+\Delta v) + \Delta m(v-w_A)$$

resp. $$m\Delta v = \Delta m w_A + \Delta m \Delta v \ .$$

Dividing by Δt and letting Δt tend to 0 gives

$$m\dot{v} = - \dot{m} w_A = u.$$

The thrust u, that is the product of the escape velocity with the amount of mass escaping per unit time, can be controlled by the rocket engines. However there are also resisting forces like air friction and gravity. Thus we get an equation

$$m\dot{v} = u - W(h,v) - mg(h),$$

where in general

$$W(h,v) = K(h)v^2 \quad (\text{e.g. } K(h) = e^{-\alpha h}),$$

$$g(h) = g_O \frac{r_o^2}{(r_o+h)^2}$$

(r_o = radius of the earth, g_O = gravitational acceleration at the surface of the earth). In addition we have the equations

$$\dot{m} = - \frac{u}{w_A}$$

$$\dot{h} = v$$

and the initial conditions $h(0) = 0$, $v(0) = 0$, $m(0) = m_o$. For the time t_1 which the rocket needs to reach its maximum altitude we have the equations $v(t_1) = 0$, $m(t_1) = m_1$ ($m_o - m_1$ is the weight of the fuel). For technical reasons one makes the restriction

$$0 \leq u \leq u_{max}$$

on the controllable thrust.

Altogether then we have the following mathematical problem:

let $u = u(t)$ be a sufficiently smooth control function, e.g. piecewise continuous with $0 \leq u(t) \leq u_{max}$. Let $h(t)$, $v(t)$, and $m(t)$ be the altitude, velocity and mass of the rocket given by the system of differential equations

$$\dot{h} = v$$

$$\dot{v} = \frac{u-W(h,v)}{m} - g(h)$$

$$\dot{m} = -\frac{u}{w_A}$$

with initial conditions $h(0) = 0$, $v(0) = 0$, $m(0) = m_o$. Let t_1 be the time required to reach the maximum altitude with a given control function u, i.e. $v(t_1) = 0$, $m(t_1) = m_1$. Among all possible control functions find the one that maximizes $h(t_1)$.

2) We consider a forced oscillation, which can be given by a linear differential equation of second order:

$$(*) \qquad \ddot{x} + a(t)\dot{x} + b(t)x = u(t).$$

This process is controlled by the external force u; therefore u is called the control function. Let the initial condition of the system be determined by

$$(**) \qquad x(0) = x_o, \quad \dot{x}(0) = 0.$$

Let functions a,b be given on a time interval [0,T] and e.g. continuous. The linear initial value problem is uniquely solvable for any sufficiently smooth u - say u piecewise continuous on [0,T]; let $x(t) = x(t;u)$ be the solution. The process shall be so controlled by appropriate choice of u that

$$x^2(T;u) + \dot{x}^2(T;u)$$

is minimal, that is so that the trajectory x is as close as possible to the origin (0,0) in the phase plane $(x,\dot{x})$ at time

T. Here it is reasonable to assume that the control function u cannot take on arbitrarily large or arbitrarily small values, i.e. that (e.g. for technical reasons) only such control functions are feasible for which $|u(t)| \le c$ for $t \in [0,T]$ with a prescribed constant c .

If we write this as an optimization problem, then we have

$$M := \{u : [0,T] \to \mathbb{R} \text{ piecewise continuous: } |u(t)| \le c \text{ for } t \in [0,T]\}$$

as the set of feasible control functions and as cost function to be minimized on M

$$C(u) := x^2(T;u) + \dot{x}^2(T;u),$$

where $x(\cdot;u)$ is the solution of the initial value problem (*), (**) for given $u \in M$.

1.6. Survey

In this section we shall try to give a short survey of the following chapters and show how the individual parts are related.

We are concerned here with the theory of optimization in the sense that no numerical methods are treated, but we should emphasize that many of the results presented here (e.g. necessary and sufficient optimality conditions) are fundamental for algorithms that are applied in practice to the numerical solution of optimization problems. The most important questions in optimization are:

- Under what assumptions do solutions exists?
- Which conditions are necessary for the existence of a solution? Do these conditions characterize a solution, i.e. are they also sufficient?
- How do the value and the solution set of an optimization problem change under perturbation of the data, that is of the objective function and the set of feasible solutions?

Unfortunately for reasons of time and space we shall not be able to treat this last question, but we shall treat the other two quite thoroughly.

We begin in § 2 with the classical duality theory for finite dimensional linear programming problems. To the original problem we associate a so-called dual program. This makes it possible to give necessary and sufficient existence and optimality conditions. In order to be able in § 4 to treat infinite dimensional convex programs we set up the necessary functional analytical tools in § 3. These are in particular separation theorems for convex sets in linear and normed linear spaces. We assume almost no knowledge of functional analysis. In § 5 we prove necessary optimality conditions for nonconvex but differentiable optimization problems in normed linear spaces. To a given starting problem one associates a linearized program and using the duality theory for convex programming from § 4 and an extra condition, a so-called constraint qualification, one proves the existence of a solution to the dual problem associated to this linearized program, that is of so-called Lagrange multipliers. Finally in § 6 we consider variations on the theorem of WEIERSTRASS that a continuous real-valued function on a compact set assumes its extreme values. In each section the abstract theorems will be made concrete by many examples and applications. By this means and by emphasizing the geometric background we shall try to give the reader an understanding of the material that goes beyond the merely formal.

1.7 Literature

1.1: One finds a survey of optimization problems in elementary geometry in ZACHARIAS [80], STURM [72]. Historical remarks on the FERMAT problem are to be found in KUHN [47]. A good survey of the significance of the FERMAT-WEBER problem in location theory is BLOECH [7].

1.2: For a very good short introduction to the history and objectives of the calculus of variations we refer the reader to the introduction of BLANCHARD-BRUENING [5], for a more extensive presentation to GOLDSTINE [29]. A relatively elementary and very readable book about the calculus of variations with many historical remarks (in particular about the brachystochrone problem) and examples is SMITH [69].

1.3.: GOLDSTINE [28] makes historical remarks about the work of GAUSS and LEGENDRE on the method of least squares. Further literature about investigations on the choice of starting approximation for the computation of square roots can be found in an essay by MEINARDUS-TAYLOR [60].

1.4.: We refer to DANTZIG [20] for interesting remarks on the history of linear programming.

1.5: For further examples of optimal control problems see e.g. BRYSON-HO [11] and KNOWLES [43].

§ 2 LINEAR PROGRAMMING

2.1 Definition and interpretation of the dual program

One says that a linear program is in normal form if all restrictions are in the form of equations and all variables are restricted in sign, i.e. if it has the form

(P) Minimize $c^T x$ on $M := \{x \in \mathbb{R}^n : Ax = b,\ x \geq 0\}$

with $c \in \mathbb{R}^n$, $b \in \mathbb{R}^m$ and $A \in \mathbb{R}^{m\times n}$. Every linear program can be transformed into an equivalent one in normal form. For if one starts with:

$$\text{minimize } c_1^T x_1 + c_2^T x_2 \quad \text{subject to}$$

$$\left.\begin{array}{l} A_{11}x_1 + A_{12}x_2 = b_1 \\ A_{21}x_1 + A_{22}x_2 \geq b_2 \end{array}\right. , \quad x_1 \geq 0,$$

then the inequality restriction $A_{21}x_1 + A_{22}x_2 \geq b_2$ is equivalent after introduction of the slack variables z to

$$A_{21}x_1 + A_{22}x_2 - z = b_2 \ , \ z \geq 0,$$

and the free variable x_2 can be regarded as the difference of two variables $x_2^+, x_2^- \geq 0$ restricted in sign: $x_2 = x_2^+ - x_2^-$. Written in normal form then the general linear program reads as follows:

$$\text{minimize} \begin{pmatrix} c_1 \\ c_2 \\ -c_2 \\ 0 \end{pmatrix}^T \begin{pmatrix} x_1 \\ x_2^+ \\ x_2^- \\ z \end{pmatrix} \quad \text{subject to}$$

$$\begin{pmatrix} A_{11} & A_{12} & -A_{12} & 0 \\ A_{21} & A_{22} & -A_{22} & -I \end{pmatrix} \begin{pmatrix} x_1 \\ x_2^+ \\ x_2^- \\ z \end{pmatrix} = \begin{pmatrix} b_1 \\ b_2 \end{pmatrix}, \quad \begin{pmatrix} x_1 \\ x_2^+ \\ x_2^- \\ z \end{pmatrix} \geq 0$$

To a linear program in normal form

(P) $\quad$ Minimize $c^T x$ on $M := \{x \in \mathbb{R}^n : Ax = b,\ x \geq 0\}$

we associate the dual program

(D) $\quad$ Maximize $b^T y$ on $N := \{y \in \mathbb{R}^m : A^T y \leq c\}$.

Remarks: 1. If one starts with the program

$$\text{Minimize } c_1^T x_1 + c_2^T x_2 \text{ subject to}$$

$$\begin{matrix} A_{11}x_1 + A_{12}x_2 = b_1 \\ A_{21}x_1 + A_{22}x_2 \geq b_2 \end{matrix} \quad , \; x_1 \geq 0.$$

then after writing this program in normal form one gets the dual program:

$$\text{Maximize } b_1^T y_1 + b_2^T x_2 \text{ subject to}$$

$$\begin{pmatrix} A_{11}^T & A_{21}^T \\ A_{12}^T & A_{22}^T \\ -A_{12}^T & -A_{22}^T \\ 0 & -I \end{pmatrix} \begin{pmatrix} y_1 \\ y_2 \end{pmatrix} \leq \begin{pmatrix} c_1 \\ c_2 \\ -c_2 \\ 0 \end{pmatrix} \quad \text{resp.} \quad \begin{matrix} A_{11}^T y_1 + A_{21}^T y_2 \leq c_1 \\ A_{12}^T y_1 + A_{22}^T y_2 = c_2 \end{matrix} \; , \; y_2 \geq 0$$

Thus one obtains the dual program to a given so-called primal linear program by first writing the primal program in normal form (P) and then associating the dual program (D) to (P).

2. If one writes (D) in normal form, one has

$$\text{Minimize } \begin{pmatrix} -b \\ b \\ 0 \end{pmatrix}^T \begin{pmatrix} y^+ \\ y^- \\ z \end{pmatrix} \quad \text{subject to}$$

$$(-A^T \quad A^T \quad -I)\begin{pmatrix} y^+ \\ y^- \\ z \end{pmatrix} = -c, \quad \begin{pmatrix} y^+ \\ y^- \\ z \end{pmatrix} \geq 0$$

The program dual to this one is

Maximize $(-c)^T x$ subject to

$$\begin{pmatrix} -A \\ A \\ -I \end{pmatrix} x \leq \begin{pmatrix} -b \\ b \\ 0 \end{pmatrix} \qquad \text{resp. } Ax = b,\ x \geq 0.$$

The program dual to (D) is thus (P) again.

Up to now the definition of a dual program has been rather formal; thus an economic interpretation (applied to the production planning and transportation problems) and a geometric interpretation may help the reader understand what is meant.

Examples: 1. In Example 2) in 1.4 we described the production planning problem. It was:

Maximize $p^T x$ subject to $Ax \leq b$, $x \geq 0$. A "contractor" offers the firm to rent or buy all resources and offers $y_i \geq 0$ units of money per unit of the ith resource; all together then his costs will be

$$b^T y = \sum_{i=1}^{m} b_i y_i$$

and he will seek to minimize this sum. The firm, however, is only willing to accept the offer if

$$\sum_{i=1}^{m} a_{ij} y_i \geq p_j \qquad (j=1,\ldots,n),$$

i.e. if the value of all resources needed to produce one unit of the jth product is no less than the net profit p_j which the firm would have earned by producing the product itself. The "contractor" thus has to solve the linear program

Minimize $b^T y$ subject to $A^T y \geq p$, $y \geq 0$, and this is precisely the dual program.

2. The transportation problem (c.f. Example 3) in 1.4) is:

$$\text{Minimize} \sum_{i=1}^{m} \sum_{j=1}^{n} c_{ij}x_{ij} \quad \text{subject to}$$

$$\sum_{j=1}^{n} x_{ij} \le r_i, \ \sum_{i=1}^{m} x_{ij} \ge t_j, \ x_{ij} \ge 0$$

$$(i=1,\ldots,m, j=1,\ldots,n).$$

A transport firm approaches the oil company to whom the refineries $R_1,\ldots,R_m$ and the tank depots $T_1,\ldots,T_n$ belong with the following offer: "We will buy the entire production r_i of R_i at a price of u_i per unit and supply T_j in such a way that its requirement of t_j units is met; we will charge v_j per unit." Thus the profit of the transport firm is

$$\sum_{j=1}^{n} t_j v_j - \sum_{i=1}^{m} r_i u_i,$$

and the firm will want to maximize this figure. But the oil company will only accept the offer if $v_j - u_i \le c_{ij}$, i.e. if the transportation of one unit from R_i to T_j is not more expensive than c_{ij}. The transport firm thus has to solve the following problem:

$$\text{Maximize} \sum_{j=1}^{n} t_j v_j - \sum_{i=1}^{m} r_i u_i \quad \text{subject to}$$

$$v_j - u_i \le c_{ij}$$

$$(i=1,\ldots,m, \ j=1,\ldots,n)$$

$$v_j, u_i \ge 0$$

and this is, as one can readily check, precisely the dual program to the transportation problem.

Now we shall give a geometric interpretation of the dual problem. Again we begin with the primal problem in normal form:

(P) $\qquad$ Minimize $c^T x$ on $M := \{x \in \mathbb{R}^n : Ax = b, \ x \ge 0\}$.

Let us define the set

$$\Lambda := \{(b-Ax, c^T x) \in \mathbb{R}^m \times \mathbb{R} : x \geq 0\}.$$

Λ obviously has the following properties:

1. The line segment joining any two points of Λ also lies entirely in Λ, i.e. Λ is convex. For if $x_1, x_2 \geq 0$, then $(b-Ax_1, c^T x_1)$, $(b-Ax_2, c^T x_2) \in \Lambda$ and for any $\lambda \in [0,1]$ we have

$$(1-\lambda)(b-Ax_1, c^T x_1) + \lambda(b-Ax_2, c^T x_2) = (b-Ax, c^T x) \in \Lambda$$

$$\text{with } x = (1-\lambda)x_1 + \lambda x_2 \geq 0.$$

2. The entire half-line from $(b,0)$ through any given point in Λ also lies in Λ (i.e. Λ is a "cone with vertex at $(b,0)$"). For if $x \geq 0$, then $(b-Ax, c^T x) \in \Lambda$ and for $\lambda \geq 0$ we have

$$(1-\lambda)(b,0) + \lambda(b-Ax, c^T x) = (b-A(\lambda x), c^T(\lambda x)) \in \Lambda.$$

Thus Λ looks roughly as follows

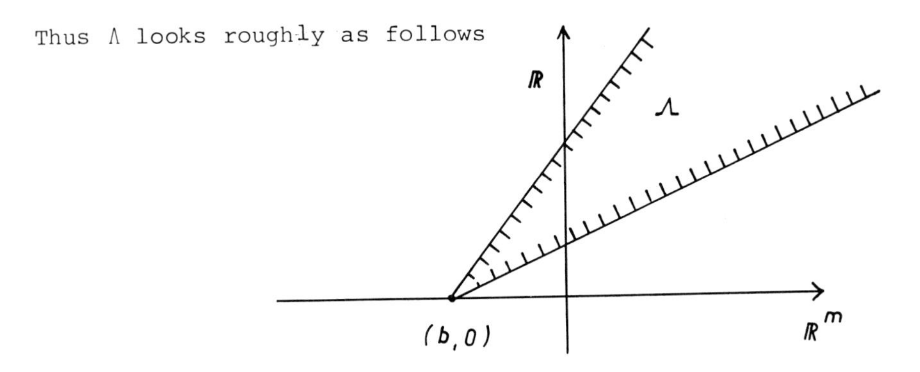

Hence we can formulate the primal problem in the following terms:

$(\tilde{P})$ Minimize β subject to $(0,\beta) \in \Lambda$.

The value inf (P) of (P) resp. $(\tilde{P})$ is therefore geometrically the lowest intersection point of Λ and the $\mathbb{R}$-axis .

For given fixed $(y,\alpha) \in \mathbb{R}^m \times \mathbb{R}$

$$H(y,\alpha) := \{(z,r) \in \mathbb{R}^m \times \mathbb{R} : r + y^T z = \alpha\}$$

is a hyperplane in $\mathbb{R}^m \times \mathbb{R}$, which is not parallel to $\mathbb{R}$.

$$H^+(y,\alpha) := \{(z,r) \in \mathbb{R}^m \times \mathbb{R} : r + y^T z \geq \alpha\}$$

is the closed nonnegative halfspace generated by this hyperplane.

To $(\tilde{P})$ we now associate the following problem:

$(\tilde{D})$ Maximize α on $\tilde{N} := \{(y,\alpha) \in \mathbb{R}^m \times \mathbb{R} : \Lambda \subset H^+(y,\alpha)\}$.

Formulated in words:

Among all hyperplanes not parallel to $\mathbb{R}$ and containing Λ in the nonnegative closed halfspace they generate find the one whose intersection point with the $\mathbb{R}$-axis is as large as possible.

The following lemma makes clear the connection between $(\tilde{D})$ and the dual program defined above

(D) Maximize $b^T y$ on $N := \{y \in \mathbb{R}^m : A^T y \leq c\}$.

<u>2.1.1 Lemma</u>: (D) and $(\tilde{D})$ are equivalent. More precisely:

i) $(y,\alpha) \in \tilde{N} \Leftrightarrow y \in N,\ \alpha \leq b^T y$.

ii) If $(y,\alpha) \in \tilde{N}$ is a solution of $(\tilde{D})$, then y is a solution of (D).

iii) If $y \in N$ is a solution of (D), then $(y, b^T y)$ is a solution of $(\tilde{D})$.

<u>Proof</u>: i) $(y,\alpha) \in \tilde{N} \Leftrightarrow c^T x + y^T(b-Ax) \geq \alpha$ for all $x \geq 0$

$$\Leftrightarrow (c-A^T y)^T x \geq \alpha - b^T y \text{ for all } x \geq 0$$

$$\Leftrightarrow A^T y \leq c,\ \alpha \leq b^T y$$

$$\Leftrightarrow y \in N,\ \alpha \leq b^T y.$$

ii), iii) follow immediately from i).

From now on we shall call an optimization problem <u>feasible</u> if the set of feasible solutions is nonempty. If both the primal program (P) and the dual program (D) are feasible, then the geometric interpretation of the two problems looks roughly as follows:

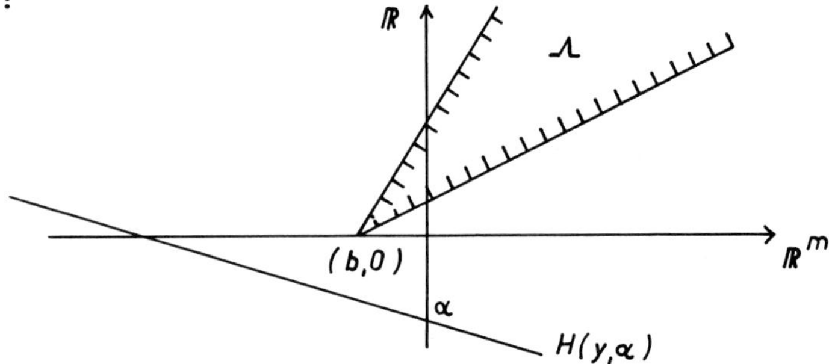

From the picture one can see that the intersection point α of a hyperplane $H(y,\alpha)$ which is feasible for the dual problem $(\tilde{D})$ is less than or equal to β for any $(0,\beta) \in \Lambda$. This geometrically obvious statement is the content of

<u>2.1.2 Theorem</u>: (Weak Duality Theorem): Suppose (P) and (D) are feasible. If $x \in M$ and $y \in N$ then $b^T y \leq c^T x$. If in addition $b^T y = c^T x$, then x is a solution of (P) and y a solution of (D).

<u>Proof</u>: We have

$$
\begin{aligned}
b^T y = (Ax)^T y = x^T A^T y \leq x^T c \qquad &(\text{because } x \geq 0) \\
= c^T x. &
\end{aligned}
$$

The rest is obvious.

If one defines the values inf(P) resp. sup (D) of the program (P) resp. (D) by

$$
\inf (P) := \begin{cases} \inf \{c^T x : x \in M\} & \text{if } M \neq \emptyset \\ +\infty & \text{if } M = \emptyset \end{cases}
$$

resp.

$$
\sup (D) := \begin{cases} \sup \{b^T y : y \in N\} & \text{if } N \neq \emptyset \\ -\infty & \text{if } N = \emptyset, \end{cases}
$$

then as an immediate consequence of the weak duality theorem one has:

i) sup (D) $\le$ inf (P)

ii) (D) feasible $\Rightarrow$ inf (P) $> -\infty$.

iii) (P) feasible $\Rightarrow$ sup (D) $< +\infty$.

Remarks: 1. It is possible with the help of the weak duality theorem to give lower bounds for the value inf (P) of (P).

2. The weak duality theorem gives a sufficient condition for an $\bar{x} \in M$ to be solution of the program (P). If there is a $\bar{y}$ with $A^T\bar{y} \le c$ and $b^T\bar{y} = c^T\bar{x}$ resp. $(c-A^T\bar{y})^T\bar{x} = 0$, then $\bar{x}$ is a solution of (P). Later we shall see that the converse is also true.

3. Earlier we gave an economic interpretation of the program dual to the transportation problem. The weak duality theorem says the profit of the transportation firm cannot be larger than the transportation costs of the oil producer.

2.2 The FARKAS-Lemma and the Theorem of CARATHEODORY

From linear algebra it is well known that the system of linear equations

$$Bx = d \quad \text{with} \quad B \in \mathbb{R}^{k\times n} \quad \text{and} \quad d \in \mathbb{R}^k$$

is solvable if and only if $d^Tz = 0$ for all z with $B^Tz = 0$. FARKAS (1902) gave a necessary and sufficient condition for a system of equations $Bx = d$ to have a nonnegative solution. This lemma turned out to be the main tool for proving the strong duality theorem, which essentially says that the feasibility of (P) and (D) implies the solvability of (P) and (D) and min (P) = max (D).

2.2.1 Lemma (FARKAS): Suppose $B \in \mathbb{R}^{k\times n}$, $d \in \mathbb{R}^k$. Then exactly one of the following alternatives holds:

i) $Bx = d$, $x \ge 0$ is solvable with an $x \in \mathbb{R}^n$.

ii) $B^Tz \ge 0$, $d^Tz < 0$ is solvable with a $z \in \mathbb{R}^k$

Before we present the formal proof of this lemma, we wish to see what the FARKAS-Lemma means geometrically.

If i) is false, then $d \notin K := \{Bx : x \geq 0\}$. The FARKAS-Lemma says that ii) must then be true, i.e. that there is a z ($\neq 0$) with $B^T z \geq 0$, $d^T z < 0$. With this z we define the hyperplane

$$H := \{y \in \mathbb{R}^k : z^T y = 0\}$$

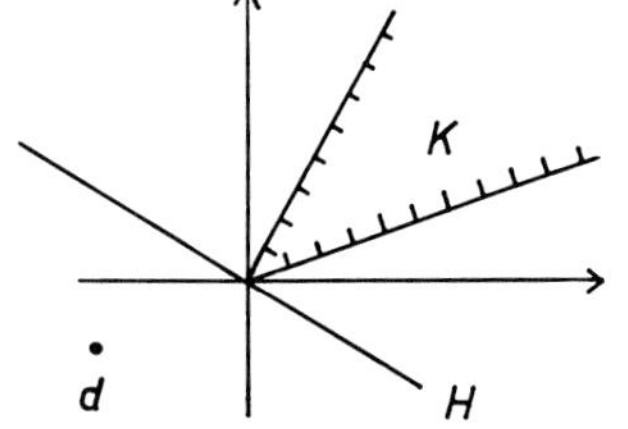

passing through the origin. Let

$$H^+ := \{y \in \mathbb{R}^k : z^T y \geq 0\}$$

be the nonnegative halfspace generated by H. Then we have

a) $K \subset H^+$, for $x \geq 0$ implies $z^T(Bx) = (B^T z)^T x \geq 0$.

b) $d \notin H^+$, for $z^T d < 0$.

In other words: if $d \notin K$, then there exists a hyperplane H whose nonnegative halfspace contains K but not d - i.e. which strictly separates d and K.

Proof of Lemma 2.2.1: a) i) and ii) are not both true! For suppose they were. Then we would have

$$0 > d^T z = (Bx)^T z = x^T B^T z \geq 0, \text{ a contradiction.}$$

b) Suppose i) is false, i.e. $Bx = d$ has no nonnegative solution. We distinguish two cases:

1. $Bx = d$ has no solution at all.
According to the theorem of linear algebra mentioned above there then exists a $z \in \mathbb{R}^k$ with $B^T z = 0$ and $d^T z \neq 0$, say $d^T z < 0$.

2. $Bx = d$ has a solution, but no nonnegative ones.
Let $B = (b^1\ b^2\ \dots\ b^n)$, i.e. b^i the ith column of B. We continue the proof with an induction on n. For $n = 1$
$Bx = x_1 b^1 = d$ only a solution $x_1 < 0$ exists. If we set $z = -d$, then

$$B^T z = b^{1T} z = -\frac{d^T d}{x_1} > 0 \text{ and } d^T z = -d^T d < 0.$$

Now we assume the statement is true for matrices with at most n - 1 columns and must show that it is also true for $B = (b^1 \ldots b^{n-1} b^n)$. Suppose

$$Bx = \sum_{j=1}^{n} x_j b^j = d$$

has no nonnegative solution. Then the same is true for

$$\sum_{j=1}^{n-1} x_j b^j = d,$$

so by the induction hypotheses there must exist a $z^* \in \mathbb{R}^k$ with $b^{jT} z^* \geq 0$ $(j=1,\ldots,n-1)$ and $d^T z^* < 0$. We may assume $b^{nT} z^* < 0$, for otherwise we would be finished. Let us set

$$\bar{b}^j := (b^{jT} z^*) b^n - (b^{nT} z^*) b^j \quad (j=1,\ldots,n-1)$$

$$\bar{d} := (d^T z^*) b^n - (b^{nT} z^*) d.$$

The equation

$$\sum_{j=1}^{n-1} \bar{x}_j \bar{b}^j = \bar{d}$$

has no nonnegative solution, for otherwise the equation

$$-\frac{1}{b^{nT} z^*} \left\{ \sum_{j=1}^{n-1} \bar{x}_j (b^{jT} z^*) - d^T z^* \right\} b^n + \sum_{j=1}^{n-1} \bar{x}_j b^j = d$$

would have a nonnegative solution in contradiction to our assumption. By the induction hypothesis there is therefore a $\bar{z} \in \mathbb{R}^k$ with

$$\bar{b}^{jT} \bar{z} \geq 0 \quad (j=1,\ldots,n-1) \quad \text{and} \quad \bar{d}^T \bar{z} < 0.$$

If we set $z := (b^{nT} \bar{z}) z^* - (b^{nT} z^*) \bar{z}$, then we have

$$b^{jT} z = \bar{b}^{jT} \bar{z} \geq 0 \quad \text{for } j = 1,\ldots,n-1$$

$$b^{nT} z = 0 \quad \text{and} \quad d^T z = \bar{d}^T \bar{z} < 0,$$

and have thus shown that ii) is true. Hence the FARKAS-Lemma is proven.

In the geometric interpretation of the FARKAS-Lemma we encountered the set

$$K := \{Bx : x \geq 0\}.$$

Obviously K has the property that a line segment joining any two points in K and the half-line from the origin through any point of K both lie in K. It is high time to give these properties names. So as not to have to repeat the following definitions later we postulate a real linear space E.

2.2.2 Definition: Let E be a real linear space.

i) $L \subset E$ is a linear subspace if $x,y \in L$, $\lambda,\mu \in \mathbb{R} \Rightarrow \lambda x + \mu y \in L$.

ii) $A \subset E$ is an affine manifold if $x,y \in A$, $\lambda \in \mathbb{R} \Rightarrow (1-\lambda)x + \lambda y \in A$.

iii) $C \subset E$ is convex if $x,y \in C$, $\lambda \in [0,1] \Rightarrow (1-\lambda)x + \lambda y \in C$.

iv) $K \subset E$ is a cone (with vertex at 0) if $x \in K$, $\lambda \geq 0 \Rightarrow \lambda x \in K$.

It is very simple to verify that an arbitrary intersection of linear subspaces resp. affine manifolds resp. convex sets resp. cones again has the corresponding property. Since the space E itself has all these properties, the following definition is reasonable.

2.2.3 Definition: Let E be a real linear space and $S \subset E$ a subset.

i) The smallest linear subspace of E which contains S, i.e. the intersection of all linear subspaces of E containing S, is the linear span of S and is denoted by span (S).
ii), iii), iv) Correspondingly we define the affine span, convex hull resp. convex conical hull of S as the smallest affine manifold resp. convex set resp. convex cone containing S and denote these by aff (S), co (S) resp. K(S).

Then we have:

<u>2.2.4 Lemma:</u> Let E be a real linear space and $S \subset E$. Then

i) $\text{span}(S) = \left\{ \sum_{i=1}^{m} \lambda_i x^i : x^i \in S, \lambda_i \in \mathbb{R} \ (i=1,\ldots,m), m \in \mathbb{N} \right\}$

ii) $\text{aff}(S) = \left\{ \sum_{i=1}^{m} \lambda_i x^i : x^i \in S, \lambda_i \in \mathbb{R} \ (i=1,\ldots,m) \text{ with } \sum_{i=1}^{m} \lambda_i = 1, m \in \mathbb{N} \right\}$

iii) $\text{co}(S) = \left\{ \sum_{i=1}^{m} \lambda_i x^i : x^i \in S, \lambda_i \geq 0 \ (i=1,\ldots,m) \text{ with } \sum_{i=1}^{n} \lambda_i = 1, m \in \mathbb{N} \right\}$

iv) $K(S) = \left\{ \sum_{i=1}^{n} \lambda_i x^i : x^i \in S, \lambda_i \geq 0 \ (i=1,\ldots,m), m \in \mathbb{N} \right\}$.

<u>Proof</u>: We only prove iv); the proofs of i) - iii) are similar. For brevity we write

$$K := \left\{ \sum_{i=1}^{m} \lambda_i x^i : x^i \in S, \ \lambda_i \geq 0 \ (i=1,\ldots,m), m \in \mathbb{N} \right\}.$$

Then we obviously have

a) $S \subset K$
b) K is a cone
c) K is convex.

By definition of $K(S)$ we have $K(S) \subset K$. To prove $K \subset K(S)$ we must show

$$x^i \in S, \ \lambda_i \geq 0 \ (i=1,\ldots,m) \Rightarrow \sum_{i=1}^{m} \lambda_i x^i \in K(S).$$

To this end one observes that:

m = 1: If $x^1 \in S \ (\subset K(S))$, $\lambda_1 \geq 0 \Rightarrow \lambda_1 x^1 \in K(S)$, since $K(S)$ is a cone.

m = 2: Suppose $x^1, x^2 \in S \ (\subset K(S))$, $\lambda_1, \lambda_2 \geq 0$; we may assume without restriction that $\lambda_1 + \lambda_2 > 0$. Since $K(S)$ is convex we have

$$\frac{\lambda_1}{\lambda_1+\lambda_2} x^1 + \frac{\lambda_2}{\lambda_1+\lambda_2} x^2 \in K(S),$$

and since $K(S)$ is a cone we have

$$\lambda_1 x^1 + \lambda_2 x^2 = (\lambda_1+\lambda_2)\left\{\frac{\lambda_1}{\lambda_1+\lambda_2} x^1 + \frac{\lambda_2}{\lambda_1+\lambda_2} x^2\right\} \in K(S).$$

For arbitrary m the statement follows by induction.

A connection to the FARKAS-Lemma is demonstrated by the following example.

Example: Let $b_1,\ldots,b^n \in \mathbb{R}^k$. Then

$$K(\{b^1,\ldots,b^n\}) = \left\{\sum_{i=1}^{n} x_i b^i : x_i \geq 0\right\}$$

$$= \{Bx : x \geq 0\},$$

where $B = (b^1 \ldots b^n) \in \mathbb{R}^{k\times n}$ is the matrix having b^i as ith column.

The above example suggests the following definition.

2.2.5 Definition: A convex cone $K \subset \mathbb{R}^k$ is called finitely generated if there is a matrix $B \in \mathbb{R}^{k\times n}$ with $K = \{Bx : x \geq 0\}$.

Remark: One calls a convex cone $K \subset \mathbb{R}^k$ polyhedral if there is a matrix $A \in \mathbb{R}^{k\times n}$ with $K = \{z \in \mathbb{R}^k : A^T z \leq 0\}$. Polyhedral cones are obviously finite intersections of closed halfspaces

$$\{z \in \mathbb{R}^k : a^{iT} z \leq 0\}.$$

A remarkable result of MINKOWSKI-WEYL says that a convex cone $K \subset \mathbb{R}^k$ is finitely generated precisely when it is polyhedral (c.f. STOER-WITZGALL [71, p. 55ff]).

As an application of the FARKAS-Lemma we obtain:

2.2.6 Theorem: A finitely generated convex cone $K \subset \mathbb{R}^k$ is closed.

Proof: By definition there exists a matrix $B \in \mathbb{R}^{k\times n}$ with $K = \{Bx : x \geq 0\}$. Let $\{d^i\} \subset K$ be a sequence converging to $d \in \mathbb{R}^k$. If $d \notin K$, then by the FARKAS-Lemma there exists a $z \in \mathbb{R}^k$ with $B^T z \geq 0$ and $d^T z < 0$. Now $d^i = Bx^i$ with $x^i \geq 0$ and thus $d^{iT} z = x^{iT} B^T z \geq 0$. Letting i tend to ∞ we find $d^T z \geq 0$, a contradiction.

The following theorem of CARATHEODORY (1907) plays an important rôle in many applications (we shall encounter one such in 2.4).

2.2.7 Theorem (CARATHEODORY): Suppose $S \subset \mathbb{R}^k$ and $x \in co(S)$. Then x is a convex linear combination of at most k + 1 points from S. More precisely: there exist $x^i \in S$, $\mu_i \geq 0$ $(i=1,\dots,m \leq k+1)$ with

$$\sum_{i=1}^{m} \mu_i = 1 \quad \text{and} \quad x = \sum_{i=1}^{m} \mu_i x^i.$$

Proof: By Lemma 2.2.4 iii) $x \in co(S)$ can be represented in the form

$$x = \sum_{i=1}^{m} \lambda_i x^i \text{ with } x^i \in S,\ \lambda_i \geq 0\ (i=1,\dots,m)$$

$$\text{and} \quad \sum_{i=1}^{m} \lambda_i = 1.$$

We show that if $m > k + 1$, then x can be represented as a convex linear combination of $m - 1$ points from S; the claim then follows.

We may assume $\lambda_i > 0$ $(i=1,\dots,m)$. Since $m - 1 > k$, there exist $r_1,\dots,r_{m-1} \in \mathbb{R}$, not all zero, with

$$\sum_{i=1}^{m-1} r_i (x^i - x^m) = 0.$$

If we set $r_m := -\sum_{i=1}^{m-1} r_i$, then $\sum_{i=1}^{m} r_i = 0$, $\sum_{i=1}^{m} r_i x^i = 0$.

Let us define $\alpha \in \mathbb{R}$ by

$$\frac{1}{\alpha} = \max_{i=1,\dots,m} \left(\frac{r_i}{\lambda_i}\right) = \frac{r_j}{\lambda_j}$$

and $\mu_i := \lambda_i - \alpha r_i$ for $i = 1,\dots,m$; then $\mu_i \geq 0$ $(i=1,\dots,m)$,

$$\sum_{i=1}^{m} \mu_i = 1 \quad \text{and} \quad \mu_j = 0; \quad \text{in addition}$$

$$x = \sum_{i=1}^{m} \lambda_i x^i = \sum_{i=1}^{m} \mu_i x^i + \alpha \sum_{i=1}^{m} r_i x^i$$

$$= \sum_{\substack{i=1 \\ i \neq j}}^{m} \mu_i x^i,$$

which is the desired representation of x as a convex linear combination of $m - 1$ elements of S.

2.3 The strong duality theorem of linear programming

Exactly as in 2.1 we assume we have a linear program in normal form

(P) Minimize $c^T x$ on $M := \{x \in \mathbb{R}^n : Ax = b,\ x \geq 0\}$.

for which the dual program is given by

(D) Maximize $b^T y$ on $N := \{y \in \mathbb{R}^m : A^T y \leq c\}$.

We remind the reader that we called (P) resp. (D) feasible if M resp. N is not empty and that we write min (P) instead of inf (P) and max (D) instead of sup (D) if (P) resp. (D) is solvable.

2.3.1 Theorem (Strong duality theorem): Suppose given the primal program (P) and the dual program (D). Then we have

i) (P), (D) both feasible $\Rightarrow$ (P), (D) both have a solution and max (D) = min (P).

ii) (D) feasible, (P) not feasible $\Rightarrow$ sup (D) = $+\infty$.

iii) (P) feasible, (D) not feasible $\Rightarrow$ inf (P) $= -\infty$.

Before we prove the strong duality theorem we wish to try to make its claims plausible by looking at the geometric interpretation of the primal program (P) and the dual program (D), which we gave in 2.1. Again let

$$\Lambda := \{(b-Ax, c^T x) \in \mathbb{R}^m \times \mathbb{R} : x \geq 0\}.$$

i) (P) feasible, (D) feasible: i.e. $\Lambda \cap \{0\} \times \mathbb{R} \neq \emptyset$ and there exists a nonvertical hyperplane

$$H(y,\alpha) = \{(z,r) \in \mathbb{R}^m \times \mathbb{R} : r + y^T z = \alpha\},$$

which contains Λ in the nonnegative halfspace it generates.

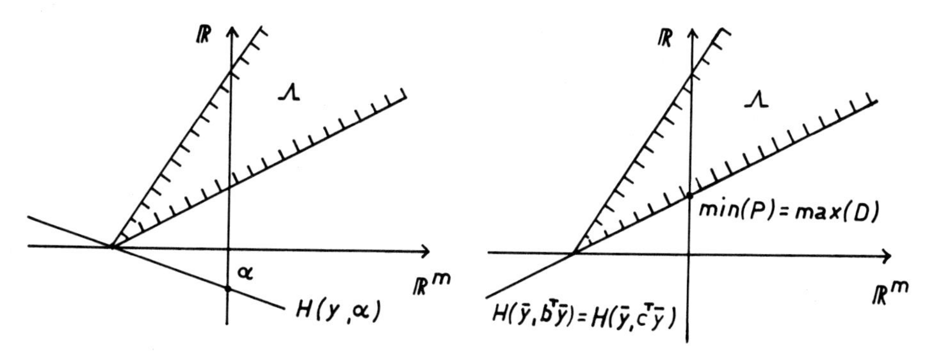

In this case it is intuitively clear that the smallest intersection point of Λ with the $\mathbb{R}$-axis and the largest possible intersection point of an admissible hyperplane with the $\mathbb{R}$-axis must agree.

ii) (D) feasible, (P) not feasible:

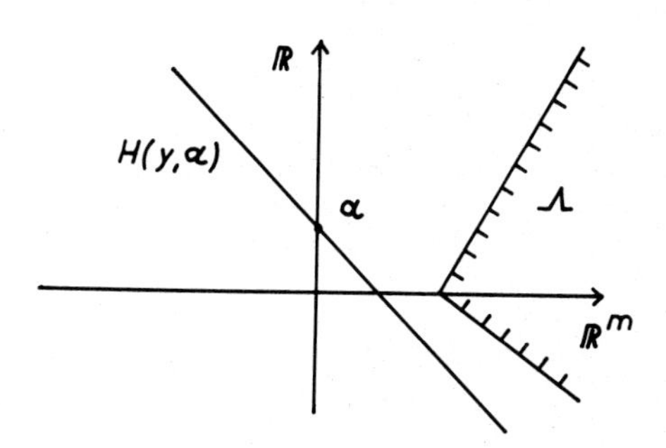

There exist hyperplanes which contain Λ in their nonnegative halfspaces and which have an arbitrarily large intersection point with the $\mathbb{R}$-axis, i.e. sup (D) $= +\infty$.

iii) (P) feasible, (D) not feasible:

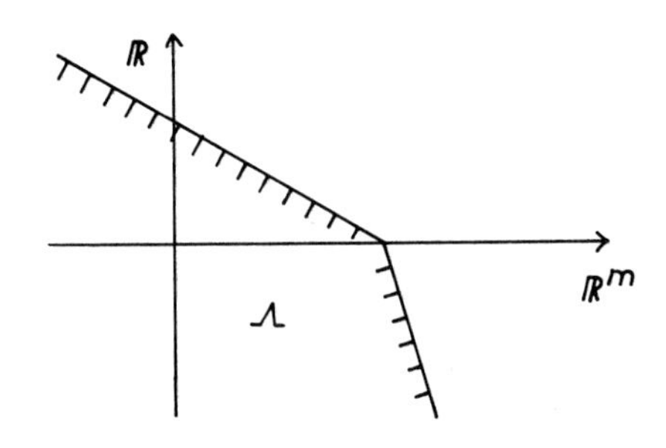

If there is no nonvertical hyperplane in $\mathbb{R}^m \times \mathbb{R}$ which contains Λ in its nonnegative halfspace and if $\Lambda \cap \{0\} \times \mathbb{R} \neq \emptyset$, then $(0,\gamma) \in \Lambda$ for all sufficiently small γ, i.e. $\inf (P) = -\infty$.

Proof of Thm. 2.3.1: i) Let (P) and (D) be feasible. From the weak duality theorem 2.1.2 it follows that $-\infty < \sup (D) < +\infty$. We demonstrate the existence of a solution of (P) and that min (P) = sup (D). The solvability of (D) then follows for reasons of symmetry, for the program dual to (D) is again (P).

We prove the existence of an $\bar{x} \in M$ with $c^T\bar{x} = \sup (D)$. Then from the weak duality theorem it follows that $\bar{x}$ solves (P) and that $c^T\bar{x} = \min (P) = \sup (D)$.

Suppose there were no $x \in M$ with $c^Tx = \sup (D)$. Then

$$\begin{pmatrix} A \\ c^T \end{pmatrix} x = \begin{pmatrix} b \\ \sup (D) \end{pmatrix}, \quad x \geq 0$$

has no solution. An application of the FARKAS-Lemma 2.2.1 provides the existence of a $(y,\gamma) \in \mathbb{R}^m \times \mathbb{R}$ with

$$(A^T\ c) \begin{pmatrix} y \\ \gamma \end{pmatrix} \geq 0, \quad \begin{pmatrix} b \\ \sup (D) \end{pmatrix}^T \begin{pmatrix} y \\ \gamma \end{pmatrix} < 0 \quad \text{resp.}$$

$$A^Ty + \gamma c \geq 0, \; b^Ty + \sup (D) \cdot \gamma < 0.$$

Because $M \neq \emptyset$ there is an $x \in \mathbb{R}^n$ with $Ax = b$, $x \geq 0$. Then we have

$$b^Ty + \gamma \sup (D) < 0 \leq (A^Ty+\gamma c)^Tx = b^Ty + \gamma c^Tx,$$

and thus $\gamma(c^Tx-\sup(D)) > 0$. From this it follows $\gamma > 0$, since by hypothesis $c^Tx > \sup(D)$ for all $x \in M$. But then we have

$$\hat{y} := -\frac{1}{\gamma}\, y \in N \quad \text{and} \quad \sup(D) < b^T\hat{y},$$

a contradiction to the definition of sup (D).

ii) Suppose (D) is feasible (say $\hat{y} \in N$) and (P) not feasible. Then $Ax = b$, $x \geq 0$ has no solution, so by the FARKAS-Lemma there is a $z \in \mathbb{R}^m$ with $A^Tz \geq 0$ and $b^Tz < 0$. If one defines $y(t) := \hat{y} - tz$, then $y(t) \in N$ for all $t \geq 0$ and

$$b^Ty(t) = b^T\hat{y} - tb^Tz \to +\infty \text{ for } t \to +\infty.$$

Thus $\sup(D) = +\infty$.

iii) Follows from ii) for reasons of symmetry.

The following theorem is an easy consequence of the strong duality theorem 2.3.1.

<u>2.3.2 Theorem</u> (Existence Theorem): Suppose given the primal program (P) and the dual program (D). Then we have: If (P) is feasible and $\inf(P) > -\infty$ or (D) is feasible and $\sup(D) < +\infty$, then (P) and (D) both have a solution and $\max(D) = \min(P)$.

<u>Proof</u>: Let (P) be feasible, $\inf(P) > -\infty$. From Theorem 2.3.1 iii) it follows that (D) is feasible and the assertion of 2.3.2 follows from Theorem 2.3.1 i). The other half of the statement follows again by symmetry.

We saw how the weak duality theorem gave a sufficient condition for an $\bar{x} \in M$ to be a solution of the primal program (P). The existence theorem and the strong duality theorem show that this condition is also necessary:

<u>2.3.3 Theorem</u>: Suppose given the linear program

(P) Minimize c^Tx on $M := \{x \in \mathbb{R}^n : Ax = b,\ x \geq 0\}$.

Then $\bar{x} \in M$ is a solution of (P) if and only if there is a $\bar{y} \in \mathbb{R}^m$ with

1. $A^T\bar{y} \le c$

2. $\bar{x}^T(c-A^T\bar{y}) = 0$ resp. $\bar{x}_j > 0 \Rightarrow c_j - (A^T\bar{y})_j = 0$ (complementary slackness condition).

Proof: i) If $\bar{x} \in M$ is a solution of (P), then $\inf (P) > -\infty$ and therefore the dual program (D)

(D) $\quad$ Maximize b^Ty on $N := \{y \in \mathbb{R}^m : A^Ty \le c\}$

has a solution $\bar{y}$ and we have $c^T\bar{x} = \min (P) = b^T\bar{y}$. Therefore

1. $A^T\bar{y} \le c$

2. $0 = c^T\bar{x} - b^T\bar{y} = \bar{x}^T(c-A^T\bar{y})$.

ii) The converse is exactly the assertion of the weak duality theorem.

The statements of the weak duality theorem 2.1.2, the strong duality theorem 2.3.1 and the existence theorem 2.3.2 obviously also hold without change for the linear program in general form

(P) $\quad$ Minimize $c_1^Tx_1 + c_2^Tx_2$ on

$$M := \left\{(x_1,x_2) \in \mathbb{R}^{n_1} \times \mathbb{R}^{n_2} : \begin{array}{l} A_{11}x_1 + A_{12}x_2 = b_1 \\ A_{21}x_1 + A_{22}x_2 \ge b_2 \end{array}, \ x_1 \ge 0\right\}$$

and the dual program

(D) $\quad$ Maximize $b_1^Ty_1 + b_2^Ty_2$ on

$$N := \left\{(y_1,y_2) \in \mathbb{R}^{m_1} \times \mathbb{R}^{m_2} : \begin{array}{l} A_{11}^Ty_1 + A_{21}^Ty_2 \le c_1 \\ A_{12}^Ty_1 + A_{22}^Ty_2 = c_2 \end{array}, \ y_2 \ge 0\right\}.$$

(The dimensions of the vectors and matrices occuring here are uniquely determined by the context.) We wish to carry Theorem 2.3.3 over to this seemingly more general case.

<u>2.3.4 Theorem</u>: Suppose given the linear program

$$\text{(P)}\qquad \begin{aligned} &\text{Minimize } c_1^T x_1 + c_2^T x_2 \text{ on} \\ &M := \{(x_1,x_2) \in \mathbb{R}^{n_1} \times \mathbb{R}^{n_2} : \begin{array}{l} A_{11}x_1 + A_{12}^T x_2 = b_1 \\ A_{21}x_1 + A_{22}x_2 \geq b_2 \end{array}, \ x_1 \geq 0\}. \end{aligned}$$

Then $(\bar{x}_1,\bar{x}_2) \in M$ is a solution of (P) if and only if there is a $(\bar{y}_1,\bar{y}_2) \in \mathbb{R}^{m_1} \times \mathbb{R}^{m_2}$ with

1. $$\begin{array}{l} A_{11}^T \bar{y}_1 + A_{21}^T \bar{y}_2 \leq c_1 \\ A_{12}^T \bar{y}_1 + A_{22}^T \bar{y}_2 = c_2, \end{array} \qquad \bar{y}_2 \geq 0$$

2. $$\bar{x}_1^T (c_1 - A_{11}^T \bar{y}_1 - A_{21}^T \bar{y}_2) = 0 \quad \text{and}$$
   $$\bar{y}_2^T (A_{21}\bar{x}_1 + A_{22}\bar{x}_2 - b_2) = 0 .$$

<u>Proof</u>: Exactly as in the proof of Theorem 2.3.3.

<u>Example</u>: Let us look at the production planning problem once more:

$$\text{Maximize } p^T x \text{ subject to } Ax \leq b,\ x \geq 0 \quad \text{resp.}$$

$$\text{Minimize } (-p)^T x \text{ subject to } (-A)x \geq -b,\ x \geq 0.$$

Applying Theorem 2.3.4 shows:

$\bar{x} \in \mathbb{R}^n$ with $A\bar{x} \leq b$, $\bar{x} \geq 0$ is a solution if and only if there is a

$\bar{y} \in \mathbb{R}^m$ with $A^T\bar{y} \geq p$, $\bar{y} \geq 0$ and $\bar{x}^T(A^T\bar{y} - p) = 0$,

$\bar{y}^T(b - A\bar{x}) = 0$.

$\bar{y}$ is of course a solution of the dual program

$$\text{Minimize } b^T y \text{ subject to } A^T y \geq p,\ y \geq 0.$$

In 2.1 we gave an interpretation of this dual program, whose solution $\bar{y}$ can be regarded as an evaluation of the m resources:

$$\bar{y}^T(b-A\bar{x}) = 0 \quad \text{resp.} \quad \bar{y}_i(b_i-(A\bar{x})_i) = 0 \ (i=1,\ldots,m)$$

then means: if in an optimal production plan the ith resource is not fully exploited, then it has value 0. Correspondingly

$$\bar{x}^T(A^T\bar{y}-p) = 0 \quad \text{resp.} \quad \bar{x}_j((A^T\bar{y})_j-p_j) = 0 \ (j=1,\ldots,n)$$

means: if the jth product is produced in an optimal production plan, then the value of all resources employed to this end is equal to the net profit p_j.

2.4 An application: relation between inradius and width of a polyhedron

The most familiar applications of the duality theory for linear programming is the fundamental theorem in the theory of two-person-zero-sum games (v. NEUMANN) and the Max-Flow/Min-Cut Theorem of network theory (FORD-FULKERSON). Presentations of these theorems are to be found in many textbooks on linear programming; we shall not go into this topic. Instead we want to take a closer look at Example 5 in 1.4. There we saw that the determination of an insphere for a given bounded polyhedron

$$P := \{x \in \mathbb{R}^n : a^{iT}x \leq b_i \ (i=1,\ldots,m)\}$$

can be reduced to the solution of a linear program. An interesting theorem of STEINHAGEN (1922) connects the inradius of a convex set in $\mathbb{R}^n$ with its width. We make the following definition (as usual $B[x;r] := \{y \in \mathbb{R}^n : |y-x| \leq r\}$ denotes the closed ball around x with radius r):

2.4.1 Definition: Suppose $P \subset \mathbb{R}^n$ is nonempty, convex and compact.

i) $\bar{r} = \bar{r}(P) := \sup \{r > 0 : \exists\, x \in P \text{ with } B[x;r] \subset P\}$ is the <u>inradius of P</u>.

ii) $\bar{w} = \bar{w}(P) := \inf\limits_{c \neq 0} \frac{1}{|c|}\{\sup\limits_{y \in P} c^T y - \inf\limits_{y \in P} c^T y\}$ is the <u>width of P</u>.

(The width of P is obviously the minimal distance between two parallel hyperplanes, one of which contains P in its nonpositive and the other in its nonnegative halfspace.)

We now show:

<u>2.4.2 Theorem</u> (STEINHAGEN): Let $a^i \in \mathbb{R}^n$, $b_i \in \mathbb{R}$ $(i=1,\dots,m)$ be given and

$$P := \{x \in \mathbb{R}^n : a^{iT}x \le b_i \ (i=1,\dots,m)\} \text{ bounded.}$$

If $\bar{r}$ is the inradius and $\bar{w}$ the width of P, then

$$\bar{w} \le \bar{r} \cdot \begin{cases} 2n^{1/2} & \text{if } n \text{ is odd} \\ \dfrac{2(n+1)}{(n+2)^{1/2}} & \text{if } n \text{ is even.} \end{cases}$$

<u>Proof</u>: We can obviously assume without restriction that $|a^i| = 1$ for $i = 1,\dots,m$. As we already saw in Example 5 of 1.4 $B[x;r] \subset P$ if and only if

$$a^{iT}x - b_i + r \le 0 \ (i=1,\dots,m), \ r \ge 0.$$

Thus the inradius $\bar{r}$ and the center $\bar{x}$ of an insphere are the solution of the linear program

$$\text{Minimize } \begin{pmatrix} -1 \\ 0 \end{pmatrix}^T \begin{pmatrix} r \\ x \end{pmatrix} \text{ subject to}$$

$$-\begin{pmatrix} 1 \\ \vdots \\ 1 \end{pmatrix} r - \begin{pmatrix} a^{1T} \\ \vdots \\ a^{mT} \end{pmatrix} x \ge -\begin{pmatrix} b_1 \\ \vdots \\ b_m \end{pmatrix}, \ r \ge 0$$

From Theorem 2.3.4 we get: there exists a $\lambda = (\lambda_i) \in \mathbb{R}^m$ with

1. $$\sum_{i=1}^{m} \lambda_i \geq 1, \quad \lambda_i \geq 0 \quad (i=1,\ldots,m)$$
$$\sum_{i=1}^{m} \lambda_i a^i = 0$$

2. $$\bar{r}\left(\sum_{i=1}^{m} \lambda_i - 1\right) = 0$$
$$\lambda_i (a^{iT}\bar{x} - b_i + \bar{r}) = 0 \qquad (i=1,\ldots,m).$$

We now distinguish two cases.

a) $\bar{r} = 0$, i.e. int $(P) = \emptyset$. We shall show that P then lies in a hyperplane and thus has width $\bar{w} = 0$, so that the statement of the theorem is trivial in this case.

Suppose aff $(P) = \mathbb{R}^n$. Then one can find $n + 1$ elements $p^o,\ldots,p^n \in P$ such that $p^1 - p^o,\ldots,p^n - p^o$ are linearly independent. Since P is convex

$$S := \text{co}(\{p^o,\ldots,p^n\}) \subset P.$$

The simplex S, however, has a nonempty interior: for example

$$\frac{1}{n+1} \sum_{j=0}^{n} p^j \in \text{int } S \subset \text{int } P, \text{ a contradiction.}$$

Thus aff $(P) \subsetneqq \mathbb{R}^n$, i.e. aff (P) and thus P are contained in a hyperplane.

b) $\bar{r} > 0$. If we define

$$\bar{I} := \{i \in \{1,\ldots,m\} : \lambda_i > 0\},$$

then $$\sum_{i\in\bar{I}} \lambda_i = 1, \quad \sum_{i\in\bar{I}} \lambda_i a^i = 0 \text{ and } \bar{r} = b_i - a^{iT}\bar{x} \text{ for } i \in \bar{I}.$$

In particular

$$0 \in \text{co}(\{a^i\}_{i\in\bar{I}}).$$

By the theorem of CARATHEODORY (Theorem 2.2.7) we can write 0 as a convex linear combination of at most n + 1 of the a^i $(i \in \bar{I})$. Thus there exists an index set $I \subset \bar{I}$ with $p := |I| \leq n + 1$ and

$$\mu_i > 0 \ (i \in I) \text{ with } \sum_{i \in I} \mu_i = 1 \text{ and } \sum_{i \in I} \mu_i a^i = 0.$$

Furthermore

$$\bar{r} = \sum_{i \in I} \mu_i \bar{r} = \sum_{i \in I} \mu_i b_i - \Big(\sum_{i \in I} \mu_i a^i\Big)^T \bar{x} = \sum_{i \in I} \mu_i b_i .$$

For $k = 1,\ldots,p-1$ let $I(k) := \{J \subset I : |J| = k\}$. If $J \in I(k)$ and $y \in P$ then

$$\sum_{j \in J} \mu_j b_j \geq \Big(\sum_{j \in J} \mu_j a^j\Big)^T y = -\Big(\sum_{j \in I \setminus J} \mu_j a^j\Big)^T y \geq -\sum_{j \in I \setminus J} \mu_j b_j$$

and thus the width of P can be estimated by

$$\bar{w} \leq \frac{\sum_{i \in I} \mu_i b_i}{|\sum_{j \in J} \mu_j a^j|} = \bar{r} \frac{1}{|\sum_{j \in J} \mu_j a^j|}$$

for all $J \in I(k)$, $k = 1,\ldots,p-1$. Thus

$$\bar{w} \leq \bar{r} \min_{k=1,\ldots,p-1} \Big(\max_{J \in I(k)} |\sum_{j \in J} \mu_j a^j|\Big)^{-1}.$$

On the other hand

$$\max_{J \in I(k)} |\sum_{j \in J} \mu_j a^j| \geq \Big\{\sum_{J \in I(k)} |\sum_{j \in J} \mu_j a^j|^2 / \binom{p}{k}\Big\}^{1/2}$$

$$\text{(because } |I(k)| = \tbinom{p}{k}\text{)}$$

$$= \Big\{\binom{p-2}{k-1} \sum_{i \in I} \mu_i^2 / \binom{p}{k}\Big\}^{1/2}$$

$$\geq \Big\{\binom{p-2}{k-1} / p \cdot \binom{p}{k}\Big\}^{1/2} = \Big\{\frac{k(p-k)}{(p-1)p^2}\Big\}^{1/2}$$

Here we have used that (remember that $|a^i| = 1$)

$$0 = \left|\sum_{i\in I} \mu_i a^i\right|^2 = \sum_{i\in I} \mu_i^2 + 2 \sum_{\substack{i,j\in I\\ i<j}} \mu_i\mu_j a^{iT}a^j$$

and therefore that

$$\sum_{J\in I(k)} \left|\sum_{j\in J} \mu_j a^j\right|^2 = \sum_{J\in I(k)} \left\{\sum_{j\in J} \mu_j^2 + 2 \sum_{\substack{i,j\in J\\ i<j}} \mu_i\mu_j a^{iT}a^j\right\}$$

$$= \binom{p-1}{k-1} \sum_{i\in I} \mu_i^2$$

$$+ \binom{p-2}{k-2} \cdot 2 \sum_{\substack{i,j\in I\\ i<j}} \mu_i\mu_j a^{iT}a^j$$

$$= \binom{p-2}{k-1} \sum_{i\in I} \mu_i^2$$

and finally, because of the CAUCHY-SCHWARZ inequality

$$1 = \sum_{i\in I} \mu_i \le p^{1/2}\left(\sum_{i\in I} \mu_i^2\right)^{1/2}$$

Altogether then

$$\bar{w} \le \bar{r} \min_{k=1,\ldots,p-1} \left\{\frac{(p-1)p^2}{k(p-k)}\right\}^{1/2}$$

and thus

$$\bar{w} \le \bar{r} \begin{cases} 2(p-1)^{1/2} & \text{if } p \text{ even} \\ \dfrac{2p}{(p+1)^{1/2}} & \text{if } p \text{ odd.} \end{cases}$$

The claim now follows easily, since $p \le n + 1$.

Remark: The upper bound in Theorem 2.4.2 is sharp. For if P is a regular simplex in $\mathbb{R}^n$, i.e.

$$P = co(\{p^o,\ldots,p^n\}) \text{ with } p^o,\ldots,p^n \in \mathbb{R}^n,\ |p^i-p^j| = d$$

constant for $0 \le i < j \le n$, then

$$\overline{w} = \overline{r} \begin{cases} 2n^{1/2} & \text{for n odd} \\ \\ \dfrac{2(n+1)}{(n+2)^{1/2}} & \text{for n even,} \end{cases}$$

as STEINHAGEN [70] showed. We leave the proof of this fact as a (not so simple) exercise.

2.5 Literature

The material in sections 2.1 to 2.3 is more or less standard and one finds it in one form or another in every book on linear programming. As representatives we mention in addition to DANTZIG [19],the virtually classical texts of GALE [25], GASS [26], HADLEY [31], and COLLATZ-WETTERLING [14].

For 2.4: STEINHAGEN [70] proves in his paper a generalization of a statement of BLASCHKE [6], who considers the 2-dimensional case.

§ 3 CONVEXITY IN LINEAR AND NORMED LINEAR SPACES

3.1 Separating convex sets in linear spaces

In this paragraph we shall answer the following question: under what conditions can one separate disjoint convex subsets A, B of a (real) linear space E by a hyperplane?

First we must explain what a hyperplane is, then what separation by a hyperplane means. We remind the reader of the definition of an affine manifold (c.f. Definition 2.2.2 ii)) $A \subset E$: A is an affine manifold in the (real) linear space E if

$$x,y \in A,\ \lambda \in \mathbb{R} \Rightarrow (1-\lambda)x + \lambda y \in A,$$

i.e. if together with any two points of A the entire line through these two points belongs to A. One can easily see that the affine manifolds are just the translates of the linear subspaces. That is

$L \subset E$ a linear subspace, $a_o \in E \Rightarrow A = a_o + L$ affine manifold

$A \subset E$ affine manifold, $a_o \in A \Rightarrow A = a_o + L$ for a linear subspace $L \subset E$ which is independent of a_o.

Here and in what follows we use the following

Notation: Suppose $A,B \in E$. Let

$$A + B := \{a + b : a \in A,\ b \in B\}.$$

If $A = \{a\}$ has only one point we also write $a + B$ instead of $\{a\} + B$. If $\Lambda \subset \mathbb{R}$ let

$$\Lambda A := \{\lambda a : \lambda \in \Lambda,\ a \in A\}.$$

For $\Lambda = \{\lambda\}$ we write λA instead of $\{\lambda\}A$.

3.1.1 Definition: i) An affine manifold $H \subset E$ is a *hyperplane* in E if H is a maximal proper affine manifold, i.e. if

1. $H \subsetneqq E$

2. If $M \subset E$ is an affine manifold with $H \subset M$, then $M = E$ or $M = H$.

ii) $E' := \{l : E \to \mathbb{R} : l \text{ linear}\}$ is the <u>algebraic dual space</u> of E. The elements of E' are called linear functionals.

If $l \in E'$ and $x \in E$, then we usually write $<l,x>$ instead of $l(x)$. If one defines addition and scalar multiplication in E' in the canonical fashion, then E' is itself a linear space.

<u>Example</u>: Let $E = \mathbb{R}^n$. If one defines $i : \mathbb{R}^n \to (\mathbb{R}^n)'$, by $i(y) := l_y$ with $<l_y,x> := y^T x$, then i is linear and bijective. $\mathbb{R}^n$ and $(\mathbb{R}^n)'$ can thus be identified with one another as linear spaces. A hyperplane H in $\mathbb{R}^n$ is of the form $H = \{x \in \mathbb{R}^n : y^T x = \gamma\}$ with $y \in \mathbb{R}^n \setminus \{0\}$, $\gamma \in \mathbb{R}$ resp. $H = \{x \in \mathbb{R}^n : <l_y,x> = \gamma\}$ with $l_y \in (\mathbb{R}^n)' \setminus \{0\}$. The next lemma says that a corresponding statement is true for arbitrary linear spaces.

<u>3.1.2 Lemma</u>: i) If $H \subset E$ is a hyperplane, then there exist $l \in E' \setminus \{0\}$ and $\gamma \in \mathbb{R}$ with $H = \{x \in E : <l,x> = \gamma\}$.

ii) If $l \in E' \setminus \{0\}$ and $\gamma \in \mathbb{R}$, then $H = \{x \in E : <l,x> = \gamma\}$ is a hyperplane in E.

<u>Proof</u>: i) Let $H = x_o + V$ for a linear subspace V. We distinguish two cases:

α) $x_o \notin V$ (H is not a hyperplane through 0). Since H is maximal, we have $\text{span}(x_o,V) = E$. Every $x \in E$ has a unique representation in the form $x = \alpha x_o + v$ with $\alpha \in \mathbb{R}$, $v \in V$. Define $l \in E' \setminus \{0\}$ by

$$<l,\alpha x_o + v> = \alpha .$$

Then we have $H = \{x \in E : <l,x> = 1\}$.

β) $x_o \in V$ ($0 \in H$, H is a hyperplane through 0). Choose $x_1 \notin V$. Then $E = \text{span}(x_1,V)$, $V = H$. Again every $x \in E$ has a unique representation in the form $x = \alpha x_1 + v$, so we define $l \in E' \setminus \{0\}$ by $<l,\alpha x_1 + v> = \alpha$. Then $H = \{x \in E : <l,x> = 0\}$.

ii) Let $V := \{x \in E : \langle l,x\rangle = 0\}$. Then V is a linear subspace of E and is in fact maximal. For if $x_o \notin V$, then $E = \mathrm{span}(x_o,V)$ since every $x \in E$ can be written in the form

$$x = \frac{\langle l,x\rangle}{\langle l,x_o\rangle} x_o + \left(x - \frac{\langle l,x\rangle}{\langle l,x_o\rangle} x_o\right) \in \mathrm{span}(x_o,V).$$

For a given $\gamma \in \mathbb{R}$ there is an $x_1 \in E$ with $\langle l,x_1\rangle = \gamma$. Thus

$$\{x \in E : \langle l,x\rangle = \gamma\} = \{x \in E : \langle l,x-x_1\rangle = 0\} = x_1 + V$$

is a hyperplane.

A hyperplane $H(l,\gamma) := \{x \in E : \langle l,x\rangle = \gamma\}$ with $l \in E' \setminus \{0\}$ and $\gamma \in \mathbb{R}$ defines two complementary halfspaces, namely

$$H^+(l,\gamma) := \{x \in E : \langle l,x\rangle \geq \gamma\}$$

and

$$H^-(l,\gamma) := \{x \in E : \langle l,x\rangle \leq \gamma\}.$$

We shall say that two subsets A,B of E are separated by the hyperplane H if A and B lie each in one of the complementary halfspaces determined by H, i.e. if $A \subset H^+$ and $B \subset H^-$ or $A \subset H^-$ and $B \subset H^+$.

We shall prove (c.f. Theorem 3.1.9) that two disjoint convex subsets of $\mathbb{R}^n$ can be separated by a hyperplane in $\mathbb{R}^n$. This assertion is not true in arbitrary linear spaces as the following example shows:

<u>Example</u> (c.f. LEMPIO [50, p. 14]): Let E be the linear space of all $x = (x_j) \in \mathbb{R}^{\mathbb{N}}$ with only finitely many nonvanishing components and A the set of all elements of E whose last non-vanishing component is positive. Then $0 \notin A$ resp. $A \cap \{0\} = \emptyset$. A and $\{0\}$ are convex. But A and $\{0\}$ cannot be separated by a hyperplane (proof?).

In infinite dimensional spaces therefore we need something more than the assumptions of disjointness and convexity of the sets A and B that are to be separated. In order to formulate this additional hypothesis we need some further notation and definitions.

Notation: For $x,y \in E$ let $[x,y] := \{(1-\lambda)x + \lambda y : \lambda \in [0,1]\}$. Moreover we let $[x,y) := \{(1-\lambda)x + \lambda y : \lambda \in [0,1)\}$. $(x,y]$ and (x,y) (for $x \neq y$) are defined correspondingly.

3.1.3 Definition: Let $A \subset E$.

i) $\text{cor}(A) := \{a \in A :$ for every $b \in E \setminus \{a\}$ there is an $x \in (a,b)$ with $[a,x] \subset A\}$
is the algebraic core of A.

ii) $\text{icr}(A) := \{a \in A :$ for every $b \in \text{aff}(A) \setminus \{a\}$ there is an $x \in (a,b)$ with $[a,x] \subset A\}$
is the relative algebraic core (or intrinsic core) of A.

iii) $\text{lin}(A) := \{a \in E :$ there is an $a_o \in A$ with $[a_o,a) \subset A\}$
is the algebraic hull of A (linearly accessible points).

Remark: If $a \in \text{cor}(A)$, then this means intuitively that one can move away from a in any direction without immediately leaving A. If A is convex then obviously

$$\text{cor}(A) = \{a \in A : \text{for every } h \in E \text{ there is a } \tau = \tau(a,h) > 0 \text{ with } a + \tau h \in A\}.$$

This follows simply from the fact that for convex A if $a, a + \tau h \in A$ then also $[a, a+\tau h] \subset A$.

In the following lemma we summarize some simple properties of these concepts. The proofs are left as exercises.

3.1.4 Lemma: Let $A \subset E$ be convex. Then we have:

i) $a_o \in \text{cor}(A),\ a \in \text{lin}(A) \Rightarrow [a_o,a) \subset \text{cor}(A)$.

ii) $\text{cor}(\text{cor}(A)) = \text{cor}(A)$

iii) cor (A) and lin (A) are convex.

iv) $\text{cor}(A) \neq \emptyset \Rightarrow \text{lin}(\text{cor}(A)) = \text{lin}(A),\ \text{cor}(\text{lin}(A)) = \text{cor}(A)$.

3.1.5 Definition: A set $A \subset E$ is finite dimensional if aff (A) is finite dimensional resp. aff (A) = x + L with L a finite dimensional linear subspace of E.

For separation theorems in finite dimensional linear spaces the following result is important.

3.1.6 Lemma: Let $A \subset E$ be nonempty, finite dimensional and convex. Then icr (A) $\neq \emptyset$.

Proof: Let $a_o \in A$ be arbitrary. Then we have

$$\text{aff } (A) = a_o + \text{span } (A-a_o)$$

(proof?). By hypothesis dim span $(A-a_o) =: n < \infty$. So one can obviously find $\{a_1,\dots,a_n\} \subset A$ with

$$\text{span } (A-a_o) = \text{span } \{a_1-a_o,\dots,a_n-a_o\}.$$

Since A is convex we have $S := \text{co } \{a_o,a_1,\dots,a_n\} \subset A$. Thus e.g. the center of gravity (barycenter)

$$a := \frac{1}{n+1} \sum_{i=0}^{n} a_i$$

is contained in icr (A). To see this, let

$$b = a_o + \sum_{i=1}^{n} \lambda_i (a_i - a_o) \in \text{aff } (A)$$

be arbitrary. We must show that $x_\lambda := a + \lambda(b-a) \in A$ for sufficiently small $\lambda > 0$. Now

$$x_\lambda = \left(\frac{1}{n+1} + \lambda\left(\frac{n}{n+1} - \sum_{i=1}^{n} \lambda_i\right)\right)a_o$$

$$+ \sum_{i=1}^{n} \left(\frac{1}{n+1} + \lambda\left(\lambda_i - \frac{1}{n+1}\right)\right)a_i \in S \subset A$$

if $\lambda > 0$ is so small that

$$\frac{1}{n+1} + \lambda\left(\frac{n}{n+1} - \sum_{i=1}^{n} \lambda_i\right) \geq 0 \text{ and } \frac{1}{n+1} + \lambda\left(\lambda_i - \frac{1}{n+1}\right) \geq 0$$

for $i = 1,\dots,n$.

As a tool for obtaining existence statements in infinite dimensional linear spaces we need the familiar

Zorn's Lemma: Let $\mathcal{C}$ be a partially ordered set (i.e. in $\mathcal{C}$ we have a relation $\leq$ with the properties

a) $x \leq x$, for all $x \in \mathcal{C}$

b) $x \leq y,\ y \leq x \Rightarrow x = y$

c) $x \leq y,\ y \leq z \Rightarrow x \leq z$),

which is inductively ordered, i.e. every totally ordered subset $\mathfrak{F} \subset \mathcal{C}$ (i.e. for all $f,g \in \mathfrak{F}$ either $f \leq g$ or $g \leq f$) has an upper bound (i.e. there exists an $x \in \mathcal{C}$ with $f \leq x$ for all $f \in \mathfrak{F}$). Then $\mathcal{C}$ has a maximal element c (i.e. $y \in \mathcal{C}$, $c \leq y \Rightarrow y = c$).

For the proof of the fundamental separation theorem we still need two lemmas.

3.1.7 Lemma: Let $A,B \subset E$ be convex and $A \cap B = \emptyset$. Then there exist convex sets $C,D \subset E$ with

1. $C \cap D = \emptyset$, $C \cup D = E$

2. $A \subset C$, $B \subset D$.

Proof: Let $\mathcal{C} := \{K \subset E : K \text{ convex}, A \subset K, K \cap B = \emptyset\}$. $\mathcal{C}$ is nonempty since $A \in \mathcal{C}$. If one introduces a partial ordering on $\mathcal{C}$ by means of the inclusion relation, then $\mathcal{C}$ is inductively ordered. For if $\mathfrak{F} \subset \mathcal{C}$ is totally ordered, then

$$K := \bigcup_{F \in \mathfrak{F}} F$$

is an upper bound for $\mathfrak{F}$. Let $C \in \mathcal{C}$ be the maximal element given by Zorn's lemma. If one defines

$$\mathcal{D} := \{K \subset E : K \text{ convex}, B \subset K, C \cap K = \emptyset\}$$

then one obtains with the same argument a maximal element $D \in \mathcal{D}$. We wish to show that C,D are the desired sets. Obviously C,D are convex, $C \cap D = \emptyset$, and $A \subset C$, $B \subset D$. It thus remains to show that $C \cup D = E$. Suppose $C \cup D \subsetneqq E$, say $x \notin C \cup D$. By the maximality of C we have

$$\emptyset \neq \text{co}\,(C \cup \{x\}) \cap B = \bigcup_{c \in C} ([c,x] \cap B).$$

Thus there exists a $d_o \in B \subset D$ and a $c_o \in C$ with $d_o \in (c_o,x)$. Likewise from the maximality of D it follows that

$$\emptyset \neq \text{co}\,(D \cup \{x\}) \cap C = \bigcup_{d \in D} ([d,x] \cap C)$$

and thus the existence of a $c_1 \in C$ and a $d_1 \in D$ with $c_1 \in (d_1,x)$. But then

$$\emptyset \neq [c_o,c_1] \cap [d_o,d_1] \subset C \cap D,$$

a contradiction.

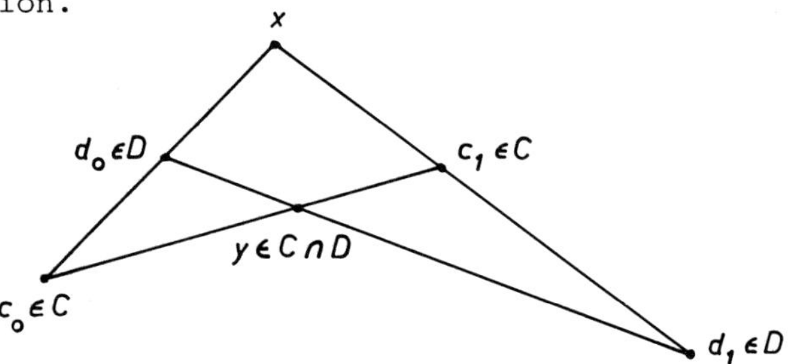

3.1.8 Lemma: Let $C,D \subset E$ be nonempty convex sets with $C \cap D = \emptyset$ and $C \cup D = E$. Let $H := \text{lin}\,(C) \cap \text{lin}\,(D)$. Then $E \setminus H = \text{cor}\,(C) \cup \text{cor}\,(D)$ and

a) $H = E$ or

b) H is a hyperplane, $\text{cor}\,(C) \neq \emptyset$ and $\text{cor}\,(D) \neq \emptyset$.

Proof: 1) $\text{lin}\,(D) = E \setminus \text{cor}\,(C)$, $\text{lin}\,(C) = E \setminus \text{cor}\,(D)$.

We leave the proof as a simple exercise.

2) $E \setminus H = \text{cor}\,(C) \cup \text{cor}\,(D)$, $H \neq \emptyset$.

From 1) we have

$$H = \text{lin}\,(C) \cap \text{lin}\,(D) = (E \setminus \text{cor}\,(C)) \cap (E \setminus \text{cor}\,(D))$$

and thus $E \setminus H = \text{cor}(C) \cup \text{cor}(D)$. Because $\emptyset \neq C \subsetneq E$ the algebraic boundary $\text{lin}(C) \setminus \text{cor}(C)$ of the convex set C is nonempty: for let $x \in C$, $z \notin C$. If one defines $\lambda_o := \sup\{\lambda \in [0,1] : x + \lambda(z-x) \in C\}$ and $y := x + \lambda_o(z-x)$, then $y \in \text{lin}(C) \setminus \text{cor}(C)$. But from 1) it follows that $H = \text{lin}(C) \setminus \text{cor}(C) = \text{lin}(D) \setminus \text{cor}(D) \neq \emptyset$.

3) H is an affine manifold in E.

By Lemma 3.1.4 iii) $\text{lin}(C)$, $\text{lin}(D)$ are convex and thus also $H = \text{lin}(C) \cap \text{lin}(D)$. Suppose H were not an affine manifold. Then there would exist $x,y \in H$ and a point z on the line through x,y which did not belong to H. Because H is convex we have $z \notin [x,y]$, say $y \in (x,z)$ and $z \in \text{cor}(C)$. By Lemma 3.1.4 i) $[z,x) \subset \text{cor}(C)$ and thus $y \in \text{cor}(C)$, a contradiction to $y \in H$.

4) $H \neq E \Rightarrow H$ is a hyperplane, $\text{cor}(C) \neq \emptyset$ and $\text{cor}(D) \neq \emptyset$.

As affine manifold H is representable in the form $H = x_o + V$ for some linear subspace V. We have to show that H resp. V is maximal. Suppose $p \notin V$; then we must show $E = \text{span}(p,V)$. From $p \notin V$ it follows $x_o + p \notin H$, thus $x_o + p \in \text{cor}(C) \cup \text{cor}(D)$. Say $x_o + p \in \text{cor}(C)$. We also have $x_o - p \notin H$, thus $x_o - p \in \text{cor}(C) \cup \text{cor}(D)$. Suppose $x_o - p \in \text{cor}(C)$. Then however $x_o \in [x_o-p, x_o+p] \subset \text{cor}(C)$, which contradicts $x_o \in H = (E \setminus \text{cor}(C)) \cap (E \setminus \text{cor}(D))$. Hence $x_o - p \in \text{cor}(D)$. In particular $\text{cor}(C)$ and $\text{cor}(D)$ are nonempty.

Let $x \in E$. We must show $x \in \text{span}(p,V)$. We distinguish two cases:

i) $x_o + x \in C$. Then $[x_o-p, x_o+x] \cap H \neq \emptyset$. For if $x_o + x \notin \text{cor}(C)$, then $x_o + x \in \text{lin}(C) \setminus \text{cor}(C) = H$, whereas if $x_o + x \in \text{cor}(C)$, then since $x_o - p \notin \text{cor}(C)$ we have the existence of a y with $x_o + y \in [x_o-p, x_o+x] \cap H$. Then $y \in V$ and

$$x_o + y = (1-\lambda)(x_o-p) + \lambda(x_o+x)$$

for some $\lambda \in (0,1]$ ($\lambda > 0$ since $p \notin V$), which implies

$$\lambda x = (1-\lambda)p + y \text{ and thus } x \in \text{span}(p,V).$$

ii) $x_o + x \notin C$ resp. $x_o + x \in D$. Then $[x_o+p, x_o+x] \cap H \neq \emptyset$ and from this one can again conclude that $x \in \text{span}(p,V)$.

Now we come to the main result of this section.

3.1.9 Theorem (Separation theorem in linear spaces):
Let A,B be disjoint nonempty convex sets in E. Suppose either cor (A) $\cup$ cor (B) $\neq \emptyset$ or E is finite dimensional. Then A and B can be separated by a hyperplane, i.e. there exist $l \in E'$, $l \neq 0$ and $\gamma \in \mathbb{R}$ with

$$\langle l,a\rangle \leq \gamma \leq \langle l,b\rangle \text{ for all } a \in A,\ b \in B.$$

Proof: By Lemma 3.1.7 there exist convex sets C,D with $A \subset C$, $B \subset D$ and $C \cap D = \emptyset$, $C \cup D = E$ (in particular C,D are non-empty). As in Lemma 3.1.8 we set $H := \text{lin}(C) \cap \text{lin}(D)$.

1) $H \subsetneq E$.
For suppose $H = E$. By Lemma 3.1.8 we then have $\emptyset = E \setminus H = \text{cor}(C) \cup \text{cor}(D)$, i.e. $\text{cor}(C) = \text{cor}(D) = \emptyset$ and $E = \text{lin}(C) = \text{lin}(D)$. We distinguish two cases.

i) cor (A) $\cup$ cor (B) $\neq \emptyset$. Say cor (A) $\neq \emptyset$. Since $A \subset C$ we have $\emptyset \neq \text{cor}(A) \subset \text{cor}(C)$, a contradiction.

ii) E finite dimensional. From $E = \text{lin}(C)$ it follows $E = \text{aff}(C)$. For if $x \in E = \text{lin}(C)$ is arbitrary, then there exists a $c_o \in C$ with $[c_o, x) \subset C$. Thus $c_1 := 1/2(c_o+x) \in C$ and $x = 2c_1 - c_o$ is an element of aff (C), so $E = \text{aff}(C)$. By Lemma 3.1.6 $\emptyset \neq \text{icr}(C) = \text{cor}(C)$, a contradiction.

2) By Lemma 3.1.8 H is a hyperplane, cor (C) and cor (D) non-empty and $E \setminus H = \text{cor}(C) \cup \text{cor}(D)$. By Lemma 3.1.2 i) there exist $l \in E'$, $l \neq 0$ and $\gamma \in \mathbb{R}$ with $H = \{x \in E : \langle l,x\rangle = \gamma\}$. Then we have $E \setminus H = \{x \in E : \langle l,x\rangle < \gamma\} \cup \{x \in E : \langle l,x\rangle > \gamma\}$. But then one of the halfspaces must coincide with cor (C), the other with cor (D) (proof?). Let's say

$$\text{cor}(C) = \{x \in E : \langle l,x\rangle < \gamma\}$$

$$\text{cor}(D) = \{x \in E : \langle l,x\rangle > \gamma\}.$$

Then we have

$$\text{lin}(\text{cor}(C)) = \text{lin}\ (C) = \{x \in E : \langle l,x\rangle \leq \gamma\}$$

and

$$\text{lin}(\text{cor}(D)) = \text{lin}\ (D) = \{x \in E : \langle l,x\rangle \geq \gamma\}$$

and since

$$A \subset C \subset \text{lin}\ (C) = \{x \in E : \langle l,x\rangle \leq \gamma\}$$

$$B \subset D \subset \text{lin}\ (D) = \{x \in E : \langle l,x\rangle \geq \gamma\}$$

the statement is proved.

Remark: In particular two disjoint nonempty convex subsets A,B of $\mathbb{R}^n$ can be separated by a hyperplane $H = \{x \in \mathbb{R}^n : y^T x = \gamma\}$ with $y \neq 0$, $\gamma \in \mathbb{R}$.

Since the condition $A \cap B = \emptyset$ in theorem 3.1.9 can be weakened, the following corollary is sometimes more applicable.

3.1.10 Corollary: Let A,B be nonempty convex subsets of E with $\text{cor}\ (A) \neq \emptyset$ and $\text{cor}\ (A) \cap B = \emptyset$. Then there exists a hyperplane $H(l,\gamma)$ ($l \in E' \setminus \{0\}$, $\gamma \in \mathbb{R}$) with

1. $\langle l,a\rangle \leq \gamma \leq \langle l,b\rangle$ for all $a \in \text{lin}\ (A)$, $b \in \text{lin}\ (B)$
2. $\langle l,a_o\rangle < \gamma$ for all $a_o \in \text{cor}\ (A)$.

Proof: Apply Theorem 3.1.9 to $\text{cor}\ (A),B$. One has $\text{cor}(\text{cor}(A)) \cup \text{cor}\ (B) = \text{cor}\ (A) \cup \text{cor}\ (B) \neq \emptyset$. Thus there exists a hyperplane $H(l,\gamma)$ with

$$\langle l,a_o\rangle \leq \gamma \leq \langle l,b\rangle \text{ for all } a_o \in \text{cor}\ (A),\ b \in B.$$

1. Let $a \in \text{lin}\ (A)$. Choose an arbitrary $a_o \in \text{cor}\ (A)$. By Lemma 3.1.4 i) we have $[a_o,a) \subset \text{cor}\ (A)$ and thus $\langle l,(1-\lambda)a_o+\lambda a\rangle \leq \gamma$ for all $\lambda \in [0,1)$. Letting λ approach 1 we obtain $\langle l,a\rangle \leq \gamma$. One shows analogously that $\gamma \leq \langle l,b\rangle$ for all $b \in \text{lin}\ (B)$.

2. Suppose there were an $a_o \in \text{cor}\ (A)$ with $\langle l,a_o\rangle = \gamma$. For

every $h \in E$ there exists a $\tau > 0$ with $a \pm \tau h \in A$. Then $\langle l,h \rangle = 0$, so $l = 0$, which contradicts the assumption that H is a hyperplane.

If one wants to separate a point from a convex set, one can get by with fewer assumptions.

3.1.11 Lemma: Let $L \subset E$ be a linear subspace and $l \in L'$. Then there exists an extension $\bar{l} \in E'$ of l : $\langle \bar{l},x \rangle = \langle l,x \rangle$ for all $x \in L$.

Proof: Let $\mathcal{C} := \{(V,f) : V$ is linear subspace of E with $L \subset V$, $f \in V'$ is an extension of $l\}$.

In $\mathcal{C}$ we define a partial ordering by the prescription

$$(V_1,f_1) \leq (V_2,f_2) \Leftrightarrow V_1 \subset V_2,\ f_2 \text{ is extension of } f_1.$$

Then $\mathcal{C}$ is inductively ordered (proof?). ZORN's Lemma assures the existence of a maximal element $(\bar{L},\bar{l})$. We show that $\bar{L} = E$. Suppose there were an $x_o \in E \setminus \bar{L}$. Let $\tilde{L} := \mathrm{span}(x_o,\bar{L})$. Every $\tilde{x} \in \tilde{L}$ has a unique representation $\tilde{x} = \alpha x_o + \bar{x}$ with $\alpha \in \mathbb{R}$, $\bar{x} \in \bar{L}$. If one defines $\tilde{l} \in \tilde{L}$ by $\langle \tilde{l},\tilde{x} \rangle = \alpha + \langle \bar{l},\bar{x} \rangle$, then $(\tilde{L},\tilde{l}) \in \mathcal{C}$ and $(\bar{L},\bar{l}) \lneq (\tilde{L},\tilde{l})$, a contradiction to the maximality of $(\bar{L},\bar{l})$.

3.1.12 Theorem: Let $A \subset E$ be convex, $\mathrm{icr}\ (A) \neq \emptyset$ and $x \notin \mathrm{icr}\ (A)$. Then there exists a hyperplane $H(\bar{l},\gamma)$ $(\bar{l} \in E' \setminus \{0\},\ \gamma \in \mathbb{R})$ with

1. $\langle \bar{l},a \rangle \leq \gamma \leq \langle \bar{l},x \rangle$ for all $a \in A$.
2. $\langle \bar{l},a \rangle < \gamma$ for all $a \in \mathrm{icr}\ (A)$.

Proof: Without restriction we may suppose $0 \in A$ (for otherwise we choose an $a_o \in A$ and set $A_o := A - a_o$. Then $A_o \subset E$ is convex, $\mathrm{icr}\ (A_o) = \mathrm{icr}\ (A) - a_o \neq \emptyset$ and $x - a_o \notin A_o$). Then $\mathrm{aff}\ (A) = \mathrm{span}\ (A) =: L$ is a linear subspace. We distinguish two cases:

a) $x \in L$. With respect to L the set A has a nonempty algebraic interior, since icr $(A) \neq \emptyset$; furthermore $\{x\} \cap \mathrm{icr}(A) = \emptyset$. From Cor. 3.1.10 we get the existence of a hyperplane $H(l,\gamma)$ in L ($l \in L' \setminus \{0\}$, $\gamma \in \mathbb{R}$) with

$$\langle l,a\rangle \leq \gamma \leq \langle l,x\rangle \quad \text{for all } a \in A \text{ and}$$

$$\langle l,a\rangle < \gamma \quad \text{for all } a \in \mathrm{icr}(A).$$

Let $\bar{l} \in E'$ be an extension of l given by Lemma 3.1.11. $H(\bar{l},\gamma)$ is then the hyperplane we require.

b) $x \notin L$. Let $\tilde{L} := \mathrm{span}(x,L)$ and define $\tilde{l} \in \tilde{L}'$ by $\langle \tilde{l},\alpha x+h\rangle := \alpha$ for $\alpha \in \mathbb{R}$, $h \in L$. Then we have $\langle \tilde{l},h\rangle = 0$ for all $h \in L$ and thus $\langle \tilde{l},a\rangle = 0$ for $a \in A$, and $\langle \tilde{l},x\rangle = 1$. Let $\bar{l} \in E'$ be an extension of $\tilde{l}$ on E given by Lemma 3.1.11. Then $H(\bar{l},1)$ is the desired hyperplane.

One should make a sketch to help unterstand the following definition.

<u>3.1.13 Definition</u>: Let $A \subset E$ be convex. A hyperplane H is called a <u>supporting hyperplane for A</u> if

1. $A \cap H \neq \emptyset$
2. A lies entirely in one of the halfspaces generated by H.

A point $a \in A$ is called a <u>support point of A</u> if there exists a supporting hyperplane H for A with $a \in A \cap H$.

From Theorem 3.1.12 we obtain immediately:

<u>3.1.14 Corollary</u>: Let $A \subset E$ be convex and icr $(A) \neq \emptyset$. Then every point $a_o \in A \setminus \mathrm{icr}(A)$ is a support point of A.

3.2 Separation of convex sets in normed linear spaces

In this paragraph $(E,\|\ \|)$ shall always denote a normed linear space over $\mathbb{R}$, i.e. E is a linear space over $\mathbb{R}$ and $\|\ \| : E \to \mathbb{R}$ is a map with

1. $\|x\| \geq 0$ for all $x \in E$; $\|x\| = 0 \Leftrightarrow x = 0$

2. $\| \alpha x \| = |\alpha| \| x \|$ for all $\alpha \in \mathbb{R}$, $x \in E$.

3. $\| x+y \| \leq \| x \| + \| y \|$ for all $x,y \in E$.

We shall use the following notation, where $x \in E$, $\varepsilon > 0$:

$$B[x;\varepsilon] := \{y \in E : \| y-x \| \leq \varepsilon\}$$

$$B(x;\varepsilon) := \{y \in E : \| y-x \| < \varepsilon\}.$$

We summarize quickly several fundamental concepts of functional analysis.

1. Let $A \subset E$. Then

$$\text{int}(A) := \{a \in A : \exists \varepsilon > 0 \text{ with } B[a,\varepsilon] \subset A\}$$

is the (topological) <u>interior of A</u>. A is <u>open</u> if $A = \text{int}(A)$ (e.g. $B(x;\varepsilon)$ is open).

2. The norm $\| \ \|$ defines a concept of <u>convergence</u> on E: a sequence $\{x_k\} \subset E$ converges to an $x \in E$ ($x_k \to x$ or $\lim_{k\to\infty} x_k = x$) if $\lim_{k\to\infty} \| x_k - x \| = 0$. The usual rules hold.

3. Let $A \subset E$. Then

$$\text{cl}(A) := \{a \in E : \exists \{a_k\} \subset A \text{ with } a_k \to a\}$$

is the (topological) <u>closure of A</u>. Obviously

$$\text{cl}(A) = \{a \in E : B[a,\varepsilon] \cap A \neq \emptyset \text{ for all } \varepsilon > 0\}.$$

A is <u>closed</u> if $A = \text{cl}(A)$ (e.g. $B[x;\varepsilon]$ is closed).

Just as with Lemma 3.1.4 we leave the proof of the following lemma as exercise for the reader.

<u>3.2.1 Lemma</u>: Let $A \subset E$ be convex. Then we have:

i) $a_o \in \text{int}(A)$, $a \in \text{cl}(A) \Rightarrow [a_o,a) \subset \text{int}(A)$.

ii) $\text{int}(A) \neq \emptyset \Rightarrow \text{cor}(A) = \text{int}(A)$, $\text{lin}(A) = \text{cl}(A)$.

iii) int (A), cl(A) are convex.

Remark: If E is finite dimensional and A convex, then cor (A) = int (A). To see this it suffices to show that $0 \in \text{cor}(A) \Rightarrow 0 \in \text{int}(A)$. We may as well assume $E = \mathbb{R}^n$. Since $0 \in \text{cor}(A)$ there is a $\tau > 0$ with $\pm \tau e^j \in A$ for $j = 1,\dots,n$, where e^j is the jth unit vector in $\mathbb{R}^n$. Then we have

$$\left\{x \in \mathbb{R}^n : |x| \le \frac{\tau}{n^{1/2}}\right\} \subset \left\{x \in \mathbb{R}^n : \| x \|_1 := \sum_{j=1}^{n} |x_j| \le \tau\right\} \subset A,$$

so $0 \in \text{int}(A)$.

On the other hand if E is infinite dimensional, then it is possible to have int (A) = $\emptyset$ and cor (A) $\neq \emptyset$ (example?).

A further very important concept is the following:

$$E^* := \left\{l \in E' : \| l \| := \sup_{x \neq 0} \frac{|\langle l,x\rangle|}{\| x \|} < \infty\right\}$$

is the (topological) <u>dual space</u> of E. One easily shows (proof?) that E^* consists precisely of the continuous linear functionals on E, i.e.

$$E^* = \{l \in E' : \{x_k\} \subset E,\ x_k \to x \Rightarrow \langle l,x_k\rangle \to \langle l,x\rangle\}$$

Remark: If E is finite dimensional, then $E^* = E'$ and E^* can be identified with E (proof?).

In general $E^* \subsetneqq E'$. In later applications we shall use the fact that one can describe explicitly the topological dual space E^* of many important normed linear spaces, which is not true of the algebraic dual E'.

In Lemma 3.1.2 we made explicit the connection between the nonzero elements of E' and the hyperplanes in E. The same connection exists between the elements of $E^* \setminus \{0\}$ and the closed hyperplanes in E, as the following lemma shows.

3.2.2 Lemma: i) If $H \subset E$ is a hyperplane, then either H = cl(H) (that is H is closed) or cl(H) = E (then one says

H is dense in E).

ii) Suppose $l \in E' \setminus \{0\}$, $\gamma \in \mathbb{R}$. Then the hyperplane $H = \{x \in E : \langle l,x \rangle = \gamma\}$ is closed if and only if $l \in E^*$.

Proof: i) Let $H = x_o + V$ be a hyperplane in E, V a linear subspace of E. Then $cl(H) = x_o + cl(V)$ is also an affine manifold (for cl(V) is a linear subspace if V is (proof?)). Furthermore $H \subset cl(H)$. Statement i) then follows from the maximality property of hyperplanes.

ii) a) Suppose $l \in E^*$. Then $H = \{x \in E : \langle l,x \rangle = \gamma\}$ is trivially closed.

b) If $H = \{x \in E : \langle l,x \rangle = \gamma\}$ is a closed hyperplane, then $V = \{x \in E : \langle l,x \rangle = 0\}$ is also closed. Choose an arbitrary $z \notin V$. Then $E = \text{span}(z,V)$ and every $x \in E$ has a unique representation $x = \alpha(x)z + v(x)$ with $\alpha(x) \in \mathbb{R}$, $v(x) \in V$. Since V is closed and $z \notin V$ the distance from z to V is positive: $d := \inf_{u \in V} \| z-u \| > 0$. Then

$$\sup_{x \neq \Theta} \frac{|\langle l,x \rangle|}{\| x \|} = \sup_{\substack{x \neq \Theta \\ x \notin V}} \frac{|\alpha(x)| \, |\langle l,z \rangle|}{|\alpha(x)| \, \| z+v(x)/\alpha(x) \|}$$

$$\leq \frac{|\langle l,z \rangle|}{d} \quad \text{since } \| z+v(x)/\alpha(x) \| \geq d$$

and hence $l \in E^*$.

If we want to show that two convex sets can be separated by a closed hyperplane under appropriate assumptions, then we shall in general apply the algebraic separation theorem and then demonstrate the continuity of the linear functional defining the hyperplane with the help of the following lemma.

3.2.3 Lemma: Let $(E, \| \ \|)$ be a normed linear space and $l \in E'$. If there exists a set $A \subset E$ with int (A) $\neq \emptyset$ and a constant $\gamma \in \mathbb{R}$ with $\langle l,x \rangle \leq \gamma$ (or $\langle l,x \rangle \geq \gamma$) for all $x \in A$, then $l \in E^*$.

Proof: Let $a_o \in$ int (A). Then there exists an $\varepsilon > 0$ with $B[a_o;\varepsilon] \subset A$. Let $x \neq 0$ be an arbitrary point of E. Then

$$a_o \pm \varepsilon \frac{x}{\| x \|} \in A, \text{ and hence } <l,a_o> \pm \varepsilon \frac{<l,x>}{\| x \|} \leq \gamma$$

and thus

$$\sup_{x \neq \Theta} \frac{|<l,x>|}{\| x \|} \leq \frac{\gamma - <l,a_o>}{\varepsilon} < \infty$$

i.e. $l \in E^*$.

Now it is simple to prove the most important separation theorems:

3.2.4 Theorem (EIDELHEIT): Let A,B be nonempty convex subsets of the normed linear space $(E, \| \ \|)$ with int (A) $\neq \emptyset$ and int (A) $\cap$ B = $\emptyset$. Then there exists a closed hyperplane $H(l,\gamma)$ ($l \in E^* \setminus \{0\}$, $\gamma \in \mathbb{R}$) with

1. $<l,a> \leq \gamma \leq <l,b>$ for all $a \in$ cl (A), $b \in$ cl (B).

2. $<l,a_o> < \gamma$ for all $a_o \in$ int (A).

Proof: By Lemma 3.2.1 ii) we have cor (A) = int (A), lin (A) = cl (A). Cor. 3.1.10 provides the existence of a hyperplane $H(l,\gamma)$ ($l \in E' \setminus \{0\}$, $\gamma \in \mathbb{R}$) with

1. $<l,a> \leq \gamma \leq <l,b>$ for all $a \in$ cl (A), $b \in$ lin (B).

2. $<l,a_o> < \gamma$ for all $a_o \in$ int (A).

From Lemma 3.2.3 it follows that $l \in E^*$. Since $\gamma \leq <l,b>$ for all $b \in B$ and $l \in E^*$ we have $\gamma \leq <l,b>$ for all $b \in$ cl (B); the theorem is proved.

An important consequence is

3.2.5 Theorem (Strict Separation Theorem): Let $(E, \| \ \|)$ be a normed linear space, $A \subset E$ convex, closed and $x \notin A$. Then $\{x\}$ and A can be strictly separated by a closed hyperplane $H(l,\gamma)$, i.e. there exist $l \in E^* \setminus \{0\}$, $\gamma \in \mathbb{R}$ with

$$<l,x> < \gamma < <l,a> \text{ for all } a \in A.$$

<u>Proof</u>: Since $x \notin \mathrm{cl}\,(A)$ we have $d := \inf_{a \in A} \| x-a \| > 0$. Thus $B(x;d) \cap A = \emptyset$. $B(x;d)$ is open and convex. Theorem 3.2.4 guarantees the existence of a closed hyperplane $H(l,\tilde{\gamma})$ with $\langle l,y \rangle < \tilde{\gamma} \le \langle l,a \rangle$ for all $y \in B(x;d)$, $a \in A$; in particular $\langle l,x \rangle < \tilde{\gamma} \le \langle l,a \rangle$ for all $a \in A$. With $\gamma := 1/2\ (\langle l,x \rangle + \tilde{\gamma})$ the claim follows.

<u>Remark</u>: Let $A,B \subset E$ be nonempty and convex, $A \cap B = \emptyset$, A closed and B <u>compact</u> (i.e. for every sequence $\{b_k\} \subset B$ there is a subsequence $\{b_{k_i}\} \subset \{b\}$ and a $b \in B$ with $\lim_{i \to \infty} b_{k_i} = b$). Then A and B can be separated strictly by a closed hyperplane, i.e. there exist $l \in E^* \setminus \{0\}$, $\gamma \in \mathbb{R}$ with $\langle l,a \rangle < \gamma < \langle l,b \rangle$ for all $a \in A$, $b \in B$. To see this note: $0 \notin A - B$, $A - B$ is nonempty, convex and closed (proof?). An application of 3.2.5 gives the result.

In 3.1.13 we defined supporting hyperplanes and support points of convex sets in linear spaces. The next theorem gives information about the existence of closed supporting hyperplanes.

<u>3.2.6 Theorem</u>: Let $(E, \| \ \|)$ be a normed linear space and $A \subset E$ nonempty and closed. Suppose $\mathrm{int}\,(A) \neq \emptyset$ or E is finite dimensional. Then every boundary point of A, that is every $a_o \in A \setminus \mathrm{int}\,(A)$, is a support point for a closed supporting hyperplane.

<u>Proof</u>: i) Suppose $\mathrm{int}\,(A) \neq \emptyset$ and $a_o \in A \setminus \mathrm{int}\,(A)$. $\{a_o\}$ and $\mathrm{int}\,(A)$ can be separated by a closed hyperplane by Theorem 3.2.4 and this is evidently the supporting hyperplane we seek.

ii) Let E be finite dimensional and $\mathrm{int}\,(A) = \emptyset$. Then $\mathrm{aff}\,(A) \subsetneqq E$ (proof?) and hence A is contained in a (closed) hyperplane.

<u>Remark</u>: In MARTI [58, p. 75ff] one finds examples which show that in Theorem 3.2.6 one cannot do without the hypothesis $\mathrm{int}\,(A) \neq \emptyset$ if E is not finite dimensional - not even if one assumes A to be compact.

<u>3.2.7 Corollary</u>: Let $(E, \|\ \|)$ be a normed linear space. Then

$$\| x \| = \max_{l \in E^*, \| l \| \leq 1} <l,x> \text{ for every } x \in E.$$

<u>Proof</u>: We may assume $x \neq 0$. Let $A := B[0; \| x \|]$. Then (by Thm. 3.2.4 or 3.2.6) there exists an $\bar{l} \in E^* \setminus \{0\}$ with $<\bar{l},a> \leq <\bar{l},x>$ for all $a \in A$. We may suppose $\| \bar{l} \| = 1$. Then

$$\| x \| \sup_{z \in B[0;1]} <\bar{l},z> = \| x \| \leq <\bar{l},x>.$$

On the other hand we trivially have

$$<l,x> \leq \| x \| \text{ for all } l \in E^* \text{ with } \| l \| = 1,$$

and the claim is proved.

3.3 Convex functions

In optimization theory convex functions play as big a role as convex sets. Since in the next chapter we want to study convex optimization problems, i.e. problems in which the objective function and the set of feasible solutions are convex, it is high time to define exactly what convex functions are and to study their properties.

<u>3.3.1 Definition</u>: Let E be a (real) linear space and $A \subset E$ convex. A real-valued function

$$f : A \to \mathbb{R}$$

is <u>convex (on A)</u> if

$$x,y \in A, t \in [0,1] \Rightarrow f(1-t)x+ty) \leq (1-t)f(x) + tf(y).$$

<u>Remarks</u>: 1. If $E = \mathbb{R}$ and $A = [a,b]$ is a compact interval, then one thinks of a convex function as looking as follows:

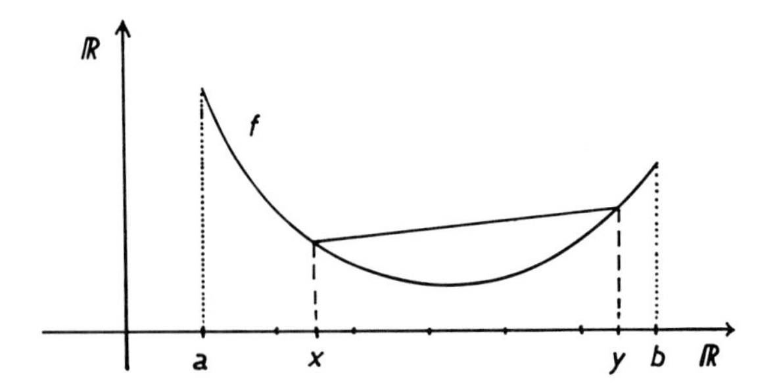

2. If $A \subset E$ is convex, then $f : A \to \mathbb{R}$ is convex if and only if the so-called epigraph of f

$$\text{epi}\,(f) := \{(x,t) \in A \times \mathbb{R} : f(x) \le t\}$$

is convex.

3. It is easy to show by induction that $f : A \to \mathbb{R}$ with A convex is convex if and only if

$$x_i \in A,\ t_i \ge 0\ (i=1,\dots,n),$$

$$\sum_{i=1}^{n} t_i = 1 \Rightarrow f\left(\sum_{i=1}^{n} t_i x_i\right) \le \sum_{i=1}^{n} t_i f(x_i).$$

Examples: 1. Suppose $E = A = \mathbb{R}^n$ and $c \in \mathbb{R}^n$ and $Q \in \mathbb{R}^{n\times n}$ is symmetric and positive semidefinite. Then $f : \mathbb{R}^n \to \mathbb{R}$ defined by

$$f(x) := c^T x + \frac{1}{2} x^T Q x$$

is convex. For if $x,y \in \mathbb{R}^n$ and $t \in [0,1]$, then

$$(1-t)f(x) + tf(y) - f((1-t)x+ty) = \frac{1}{2} t(1-t)(x-y)^T Q(x-y) \ge 0.$$

2. Let $(E, \|\ \|)$ be a normed linear space, $y \in E$, and $f : E \to \mathbb{R}$ defined by $f(x) := \| x-y \|$. Then f is convex (proof?). Is $f(x) := \| x-y \|^2$ also convex?

Notation: If E is a linear space and $A \subset E$ is convex let

$$\text{Conv}(A) := \{f : A \to \mathbb{R} : f \text{ convex}\}.$$

We shall show now that the elements of Conv (A) have some perhaps surprising smoothness properties.

3.3.2 Theorem: Let E be a linear space, $f \in$ Conv (E). Then for every $x \in E$ and for every $h \in E$ the one-sided directional derivative exists

$$f'(x;h) := \lim_{t \to 0+} \frac{f(x+th)-f(x)}{t}$$

and we have $f(x) - f(x-h) \leq f'(x;h) \leq f(x+h) - f(x)$.

Proof: Let us define $\varphi : (0,\infty) \to \mathbb{R}$ by

$$\varphi(t) := \frac{f(x+th)-f(x)}{t}$$

One shows 1) $f(x) - f(x-h) \leq \varphi(t)$ for all $t > 0$.

2) φ is nondecreasing on $(0,\infty)$ and $\varphi(s) \leq f(x+h) - f(x)$ for $s \in (0,1]$.

The assertion follows.

1) For $t > 0$:

$$f(x) = f\left(\frac{1}{1+t}(x+th) + \frac{t}{1+t}(x-h)\right)$$

$$\leq \frac{1}{1+t} f(x+th) + \frac{t}{1+t} f(x-h) \Rightarrow 1)$$

2) Let $0 < s \leq t$. Then

$$f(x+sh) - f(x) = f\left(\frac{s}{t}(x+th) + \frac{t-s}{t} x\right) - f(x)$$

$$\leq \frac{s}{t}(f(x+th)-f(x)) \Rightarrow 2)$$

The next theorem summarizes the most important properties of the map $h \to f'(x;h)$.

3.3.3 Theorem: Let E be a linear space, $f \in$ Conv (E). The map $f'(x;\cdot) : E \to \mathbb{R}$ defined by

$$f'(x;h) := \lim_{t \to 0+} \frac{f(x+th)-f(x)}{t}$$

has the following properties:

i) $f'(x;\cdot)$ is positive homogeneous, i.e.

$$f'(x;\alpha h) = \alpha f'(x;h) \text{ for all } \alpha \geq 0,\ h \in E.$$

ii) $f'(x;\cdot)$ is subadditive, i.e.

$$f'(x;h+k) \leq f'(x;h) + f'(x;k) \text{ for all } h,k \in E.$$

iii) $-f'(x;-h) \leq f'(x;h)$ for all $h \in E$.

iv) $f'(x;\cdot) \in E'$ (i.e. linear) $\Longleftrightarrow$

$$f'(x;h) = \lim_{t \to 0} \frac{f(x+th)-f(x)}{t}$$

for every $h \in E$.

<u>Proof</u>: i) is obvious.

ii) We have

$$\begin{aligned} f(x+t(h+k)) &= f(\tfrac{1}{2}(x+2th)+ \tfrac{1}{2}(x+2tk)) \\ &\leq \tfrac{1}{2}(f(x+2th)+ f(x+2tk)) \end{aligned}$$

and thus for $t > 0$

$$\frac{f(x+t(h+k))-f(x)}{t} \leq \frac{f(x+2th)-f(x)}{2t} + \frac{f(x+2tk)-f(x)}{2t}$$

We obtain the assertion by letting t approach 0 from above.

iii) $0 = f'(x;0) \leq f'(x;h) + f'(x;-h)$ by ii).

iv) is obvious by i), ii) and iii).

<u>Examples</u>: 1) Let $E = \mathbb{R}^n$ and $f(x) := c^T x + \frac{1}{2} x^T Q x$ with

$c \in \mathbb{R}^n$ and a symmetric $Q \in R^{n\times n}$. Then

$$f'(x;h) = \lim_{t\to 0+} \frac{f(x+th)-f(x)}{t}$$

$$= \lim_{t\to 0+} \{(c+Qx)^T h + \frac{1}{2}\, th^T Qh\}$$

$$= (c+Qx)^T h.$$

2) Let $E = \mathbb{R}^n$ and $f(x) := |x|$ ($|\ \ |$ = euclidean norm). Then

$$f'(x;h) = \begin{cases} \dfrac{x^T h}{|x|} & \text{for } x \neq 0 \\ |h| & \text{for } x = 0 \end{cases}$$

3) Let $E = C(B)$ be the linear space of real-valued continuous functions on the compact set $B \subset \mathbb{R}^N$. Let $z \in C(B)$ be given and

$$f(x) := \max_{t\in B} |x(t)-z(t)|.$$

Then $f \in \text{Conv}\ (E)$. We have

$$f'(x;h) = \begin{cases} \max\limits_{t\in B(x)}\ \text{sign}(x(t)-z(t))h(t) & \text{for } x \neq z \\ \max\limits_{t\in B} |h(t)| & \text{for } x = z. \end{cases}$$

where $B(x) := \{t \in B : |x(t)-z(t)| = f(x)\}$.

To prove this it evidently suffices to consider the case $x \neq z$.

a) Suppose $s_k \to 0+$. By definition of f there is a sequence $\{t_k\} \subset B$ with

$$f(x+s_k h) = |x(t_k)+s_k h(t_k)-z(t_k)|.$$

Since B is compact, $\{t_k\}$ has a convergent subsequence. We may as well assume $\{t_k\}$ itself is convergent: $t_k \to t \in B$. Obviously $t \in B(x)$ and since $x \neq z$ we have $x(t) - z(t) \neq 0$. Thus

$$\operatorname{sign}(x(t_k)+s_k h(t_k)-z(t_k)) = \operatorname{sign}(x(t_k)-z(t_k))$$

$$= \operatorname{sign}(x(t)-z(t))$$

and thus

$$\frac{f(x+s_k h)-f(x)}{s_k} \le \operatorname{sign}(x(t)-z(t))h(t_k)$$

for all sufficiently large k. Letting k go to ∞ we get

$$f'(x;h) \le \operatorname{sign}(x(t)-z(t))h(t)$$

$$\le \max_{t \in B(x)} \operatorname{sign}(x(t)-z(t))h(t).$$

b) Let $t \in B(x)$ and $s_k \to 0+$ be arbitrary. Then

$$\operatorname{sign}(x(t)+s_k h(t)-z(t)) = \operatorname{sign}(x(t)-z(t))$$

and thus

$$\frac{f(x+s_k h)-f(x)}{s_k} \ge \frac{|x(t)+s_k h(t)-z(t)|-|x(t)-z(t)|}{s_k}$$

$$= \operatorname{sign}(x(t)-z(t))h(t)$$

for all sufficiently large k. Letting k tend to ∞ we get

$$f'(x;h) \ge \operatorname{sign}(x(t)-z(t))h(t)$$

and since $t \in B(x)$ was arbitrary we also have

$$f'(x;h) \ge \max_{t \in B(x)} \operatorname{sign}(x(t)-z(t))h(t)$$

a) and b) together give the desired assertion.

In Theorem 3.3.2 we showed: if $f \in \operatorname{Conv}(E)$ then

$$f'(x;h) \le f(x+h) - f(x) \quad \text{for all } x,h \in E \text{ resp.}$$

$$f'(x;z-x) \le f(z) - f(x) \quad \text{for all } x,z \in E.$$

If $f'(x;\cdot)$ is linear, i.e. an element of E', then one calls $f'(x;\cdot)$ the gradient of f at x. This suggests the following definition.

<u>3.3.4 Definition</u>: Let E be a linear space, $A \subset E$ convex and $f \in \text{Conv}(A)$. Then

$$\partial f(x) := \{l \in E' : \langle l, z-x\rangle \leq f(z) - f(x) \text{ for all } z \in A\}$$

is called the <u>subdifferential</u> of f at x; an element $l \in \partial f(x)$ is called a <u>subgradient</u> of f at x. f is called <u>subdifferentiable</u> at x if $\partial f(x) \neq \emptyset$.

The following Lemma, whose proof is trivial, gives a geometric interpretation of the subgradient resp. subdifferential.

<u>3.3.5 Lemma</u>: Let E be a linear space, $A \subset E$ convex and $f \in \text{Conv}(A)$. Then we have

$$l \in \partial f(x) \iff H := \{(z,t) \in E \times \mathbb{R} : \langle l,z\rangle - t = \langle l,x\rangle - f(x)\}$$

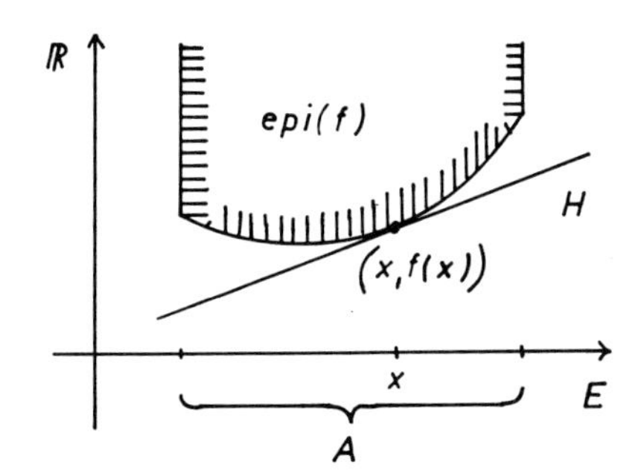

is a supporting hyperplane for the epigraph $\text{epi}(f) = \{(z,t) \in A \times \mathbb{R} : f(z) \leq t\}$ in $(x, f(x))$, i.e. $(x,f(x)) \in \text{epi}(f) \cap H$ and $\text{epi}(f) \subset H^-$.

<u>Examples</u>: 1) Let $E = \mathbb{R}$, $A = [-1,1]$ and $f(x) = |x|$. If one remembers that one can identify R' with R, then one has

$$\partial f(x) = \begin{cases} \{\text{sign}(x)\} & \text{for } x \in [-1,1] \setminus \{0\} \\ [-1,1] & \text{for } x = 0. \end{cases}$$

2) Let $E = \mathbb{R}$, $A = [-1,1]$ and $f(x) := -(1-x^2)^{1/2}$. Then $f \in \text{Conv}(A)$ and $\partial f(1) = \emptyset$ (proof?).

<u>3.3.6 Theorem:</u> Let E be a linear space, $A \subset E$ convex and $f \in \text{Conv}(A)$. Then $\partial f(x) \neq \emptyset$ for every $x \in \text{icr}(A)$, i.e. f is subdifferentiable in every point of the relative algebraic core.

<u>Proof:</u> The idea of the proof consists in showing that for every $x \in \text{icr}(A)$ there exists a nonvertical supporting hyperplane for epi (f) at $(x,f(x))$. We do this by showing that $\text{icr}(\text{epi}(f)) \neq \emptyset$ and $(x,f(x)) \notin \text{icr}(\text{epi}(f))$ and applying the separation theorem 3.1.12.

1) $\text{aff}(\text{epi}(f)) = \text{aff}(A) \times \mathbb{R}$.
For let $(x,t) \in A \times \mathbb{R}$ be arbitrary. If $f(x) \le t$ then $(x,t) \in \text{epi}(f) \subset \text{aff}(\text{epi}(f))$. If on the other hand $f(x) > t$, then

$$(x,t) = 2(x,f(x)) - (x,2f(x)-t) \in \text{aff}(\text{epi}(f)),$$

so altogether we have $A \times \mathbb{R} \subset \text{aff}(\text{epi}(f))$. But then

$$\text{aff}(A) \times \mathbb{R} = \text{aff}(A \times \mathbb{R}) \subset \text{aff}(\text{epi}(f)) \subset \text{aff}(A \times \mathbb{R}) = \text{aff}(A) \times \mathbb{R}.$$

2) Let $x \in \text{icr}(A)$. Then $(x,t) \in \text{icr}(\text{epi}(f))$ for every $t > f(x)$, and in particular $\text{icr}(\text{epi}(f)) \neq \emptyset$ and $(x,f(x)) \notin \text{icr}(\text{epi}(f))$.

For let $(y,s) \in \text{aff}(A) \times \mathbb{R} = \text{aff}(\text{epi}(f))$ be arbitrary. Since $x \in \text{icr}(A)$ there exists a $\tau_o \in (0,1)$ with $x + \tau_o(y-x) \in A$. For given $t > f(x)$ choose a $\lambda \in (0,1)$ so small that

$$(1-\lambda)f(x) + \lambda f(x+\tau_o(y-x)) \le (1-\lambda\tau_o)t + \lambda\tau_o s.$$

Then $x + \lambda\tau_o(y-x) \in A$ and

$$\begin{aligned} f(x+\lambda\tau_o(y-x)) &\le (1-\lambda)f(x) + \lambda f(x+\tau_o(y-x)) \\ &\le (1-\lambda\tau_o)t + \lambda\tau_o s, \end{aligned}$$

and thus $(x,t) + \lambda\tau_o((y,s)-(x,t)) \in epi(f) \Rightarrow (x,t) \in icr(epi(f))$. Furthermore $(x,f(x)) \notin$ **icr**$(epi(f))$, since for all $\tau \in (0,1)$ we have $(x,f(x)) + \tau((x,f(x)-1) - (x,f(x))) \notin epi(f)$.

3) An application of Theorem 3.1.12 gives the existence of an

$$(l,s) \in E' \times \mathbb{R} \setminus \{(0,0)\}, \ \gamma \in \mathbb{R} \ \text{ for } x \in icr\ (A)$$

with

$$\langle l,z\rangle - st \leq \gamma \leq \langle l,x\rangle - sf(x) \ \text{ for all } (z,t) \in epi(f)$$

$$\langle l,z\rangle - st < \gamma \qquad \text{for all } (z,t) \in icr(epi(f)).$$

Obviously $s \neq 0$. Since $(x,t) \in icr(epi(f))$ for all $t > f(x)$ we have $s > 0$. We may assume $s = 1$. From

$$\langle l,z\rangle - st \leq \langle l,x\rangle - f(x) \ \text{ for all } (z,t) \in epi(f)$$

and Lemma 3.3.5 we get the assertion.

Now we shall deal briefly with the continuity properties of convex functions. We remind the reader of the definition of a continuous function.

<u>3.3.7 Definition:</u> Let $(E_1, \|\ \|_1)$, $(E_2, \|\ \|_2)$ be normed linear spaces and $F : D \subset E_1 \to E_2$ a map with domain of definition D. F is <u>continuous</u> at $x \in D$ if for every $\varepsilon > 0$ there exists a $\delta = \delta(\varepsilon)$ with: $\| y-x \|_1 \leq \delta,\ y \in D \Rightarrow \| F(y)-F(x) \|_2 \leq \varepsilon$ (or: $\{x_k\} \subset D,\ x_k \to x \in D \Rightarrow F(x_k) \to F(x)$). F is <u>continuous on</u> $A \subset D$ if F is continuous at every $x \in A$.

If $(E, \|\ \|)$ is a normed linear space and $f : A \subset E \to \mathbb{R}$ continuous at a point $p \in A$ then for $\varepsilon = 1$ there exists a $\delta = \delta(1)$ with

$$x \in B[p;\delta] \cap A \Rightarrow f(x) \leq f(p) + 1,$$

so f is necessarily bounded above in a neighborhood of p. The next theorem gives a condition that guarantees the converse.

<u>3.3.8 Theorem:</u> Let $(E,\|\ \|)$ be a normed linear space, $A \subset E$ open and convex and $f \in \text{Conv}(A)$. If there is a sphere $B[p;r] \subset A$ with $\alpha := \sup\{f(x) : x \in B[p;r]\} < \infty$ then f is continuous on A.

<u>Proof:</u> 1) Suppose $q \in A$ is arbitrary. Then f is also bounded above on a sphere around q! Since A is open, there exists a $t > 1$ with $p + t(q-p) \in A$. Thus $B[q;(1-\frac{1}{t})r] \subset A$. For if $x = q + (1-\frac{1}{t})v$ with $v \in B[0;r]$ is an element of $B[q;(1-\frac{1}{t})r]$ then

$$x = q - (1-\frac{1}{t})p + (1-\frac{1}{t})(p+v)$$

$$= \frac{1}{t}(p+t(q-p)) + (1-\frac{1}{t})(p+v) \in A,$$

since A is convex and $B[p;r] \subset A$. Furthermore from the convexity of f we have

$$\sup\{f(x) : x \in B[q;(1-\frac{1}{t})r]\} \le \frac{1}{t} f(p+t(q-p))+(1-\frac{1}{t})\alpha.$$

2) We now demonstrate the continuity of f at p. Since we only use that f is convex and bounded above on a sphere around p, the entire claim of the theorem follows by 1) above.

We may assume $f(p) < \alpha$ (otherwise $f(x) = f(p) = \alpha$ for all $x \in B[p;r]$ and f is continuous at p).

Let $\varepsilon > 0$ be given, $t(\varepsilon) := \min\left(1,\frac{\varepsilon}{\alpha-f(p)}\right)$ and $\delta := \delta(\varepsilon) := t(\varepsilon)r$. For $x \in B[p;\delta]$ we have

$$x = (1-t(\varepsilon))p + t(\varepsilon)(p+v) \text{ with } v \in B[0;r],$$

so $f(x) \le (1-t(\varepsilon))f(p) + t(\varepsilon)\alpha$ and $f(x) - f(p) \le \varepsilon$. On the other hand

$$p = \frac{1}{1+t(\varepsilon)} x + \frac{t(\varepsilon)}{1+t(\varepsilon)} (p-v)$$

and thus $f(p) \le \frac{1}{1+t(\varepsilon)} f(x) + \frac{t(\varepsilon)}{1+t(\varepsilon)}\alpha$ resp. $-\varepsilon \le f(x) - f(p)$. Altogether then $x \in B[p;\delta] \Rightarrow |f(x)-f(p)| \le \varepsilon$ and that is the continuity of f at p.

A simple consequence of this theorem is:

3.3.9 Corollary: Let $A \subset \mathbb{R}^n$ be convex, $f \in \mathrm{Conv}(A)$. Then f is continuous on int (A).

Proof: int (A) is convex and $f \in \mathrm{Conv}(\mathrm{int}(A))$. Let $p \in \mathrm{int}(A)$. Then there exists a simplex $S = \mathrm{co}(\{a^0,\ldots,a^n\}) \subset A$ with $p \in \mathrm{int}(S)$. f is bounded above on S and hence on a sphere around p (say with respect to the euclidean norm). For if $x = \sum_{j=0}^{n} \lambda_j a^j \in S$ with $\lambda_j \geq 0$ $(j=0,\ldots,n)$, $\sum_{j=1}^{n} \lambda_j = 1$, then

$$f(x) \leq \sum_{j=0}^{n} \lambda_j f(a^j) \leq \max_{j=0,\ldots,n} f(a^j) =: \alpha .$$

The conclusion follows from Theorem 3.3.8.

Remark: In Cor. 3.3.9 we could replace int (A) by icr (A) (proof?).

The following theorem is an analog of Theorem 3.3.6 and gives us information about the existence of continuous subgradients.

3.3.10 Theorem: Let $(E, \|\ \|)$ be a normed linear space, $A \subset E$ convex with int $(A) \neq \emptyset$. Let $f \in \mathrm{Conv}(A)$ be continuous at $x \in \mathrm{int}(A)$. Then $\partial f(x) = \partial f(x) \cap E^* \neq \emptyset$.

Proof: By Theorem 3.3.6 $\partial f(x) \neq \emptyset$. We must show that $l \in \partial f(x)$ is continuous.

Since $l \in \partial f(x)$ we have $\langle l,z\rangle \leq f(z) + \langle l,x\rangle - f(x)$ for all $z \in A$. As we explained before Theorem 3.3.8 the existence of a sphere $B[x;\delta] \subset A$ on which f is bounded above follows from the continuity of f at $x \in \mathrm{int}(A)$. Therefore for $z \in B[x;\delta]$

$$\langle l,z\rangle \leq \sup\{f(z) : z \in B[x;\delta]\} + \langle l,x\rangle - f(x).$$

From Lemma 3.2.3 it follows that $l \in E^*$ and the theorem is proved.

Later we shall encounter examples for subdifferentials and

further properties of convex functions, in particular in 4.3.

We close this section with a generalization of the definition of a convex function. We remind the reader of the definition of a cone in a linear space E (c.f. Def. 2.2.2 iv)): $K \subset E$ is a cone if $x \in K$, $\lambda \geq 0 \Rightarrow \lambda x \in K$. One readily shows that a cone is convex if and only if $K + K \subset K$.

3.3.11 Definition: Let E,Z be linear spaces, $A \subset E$ convex and $K \subset Z$ a convex cone. A map $g : A \subset E \to Z$ is K-convex on A if:

$$x,y \in A,\ t \in [0,1] \Rightarrow (1-t)g(x) + tg(y) - g((1-t)x+ty) \in K.$$

If A = E we say g is K-convex.

If $Z = \mathbb{R}$ and $K = [0,\infty)$, then we shall of course simply speak of convex and not of $[0,\infty)$-convex functions.

Example: Let E be a linear space, A = E and $Z = \mathbb{R}^{m+k}$. Then

$$K := \{z \in \mathbb{R}^{m+k} : z_i \geq 0 \ (i=1,\ldots,m),\ z_i = 0 \ (i=m+1,\ldots,m+k)\}$$

is a convex cone in $\mathbb{R}^{m+k}$. For m = 0 we have $K = \{0\}$, for k = 0 K is the so-called nonnegative orthant in $\mathbb{R}^m$. A map $g : E \to \mathbb{R}^{m+k}$ with $g(x) = (g_1(x),\ldots,g_{m+k}(x))$ is K-convex if and only if $g_i \in$ Conv (E) for $i = 1,\ldots,m$ and g_i affine linear for $i = m+1,\ldots,m+k$, i.e. if $l_i \in E'$ and $\alpha_i \in \mathbb{R}$ exist with $g_i(x) = \langle l_i,x\rangle + \alpha_i$.

3.4 Literature

Separation theorems for convex sets in linear and normed linear spaces are proved in many books on functional analysis, e.g. KOETHE [44]. In our presentation and notation we have largely followed HOLMES [36].

§ 4 CONVEX OPTIMIZATION PROBLEMS

4.1 Examples of convex optimization problems

If the set M of feasible solutions of an optimization problem is a convex subset of a linear space X and the objective function $f : X \to \mathbb{R}$ is convex, then one speaks of a convex optimization problem. We shall investigate problems of the form

(P) $\quad$ Minimize $f(x)$ on $M := \{x \in X : x \in C,\ g(x) \in -K\}$

and later generally assume that $f : X \to \mathbb{R}$ is convex, $C \subset X$ is convex and $g : X \to Y$ is a map which is convex with respect to a cone K contained in the linear space Y. One easily convinces oneself that under these conditions (P) is a convex optimization problem.

Now we shall give several examples, to which we shall come back later on.

Examples: 1) The linear program

$$\text{Minimize } c^T x \text{ on } M := \{x \in \mathbb{R}^n : Ax = b,\ x \geq 0\}$$

with $c \in \mathbb{R}^n$, $b \in \mathbb{R}^m$ and $A \in \mathbb{R}^{m\times n}$ is of the type (P) if we set $X := \mathbb{R}^n$, $f(x) := c^T x$, $C := \{x \in \mathbb{R}^n : x \geq 0\}$, $Y := \mathbb{R}^m$, $K := \{0\}$ and $g(x) := b - Ax$.

2) By a quadratic optimization problem in normal form we shall mean a problem of the type

$$\text{Minimize } c^T x + \frac{1}{2} x^T Q x \text{ on } M := \{x \in \mathbb{R}^n : Ax = b,\ x \geq 0\},$$

where $Q \in \mathbb{R}^{n\times n}$ is assumed to be symmetric and positive semidefinite. In this case also the objective function $f(x) := c^T x + \frac{1}{2} x^T Q x$ is convex and the other data agree with those of 1). The general quadratic program

$$\text{Minimize } c_1^T x_1 + \frac{1}{2} x_1^T Q_1 x_1 + c_2^T x_2 + \frac{1}{2} x_2^T Q_2 x_2 \text{ subject to}$$

$$\left.\begin{array}{l} A_{11}x_1 + A_{12}x_2 = b_1 \\ A_{21}x_1 + A_{22}x_2 \geq b_2 \end{array}\right. , \quad x_1 \geq 0$$

with $c_j \in \mathbb{R}^{n_j}$, $b_j \in \mathbb{R}^{m_i}$, $A_{ij} \in \mathbb{R}^{m_i \times n_j}$ and symmetric positive semidefinite $Q_j \in \mathbb{R}^{n_j \times n_j}$ $(i,j = 1,2)$ can likewise be written as a convex program of the form (P).

3) Let $(X, \|\ \|)$ be a normed linear space, $C \subset X$ convex and $f(x) := \| x-z \|$ with fixed $z \in X$. The problem of minimizing $f(x)$ on C is called a convex approximation problem; if $C = V$ is a linear subspace one speaks of a linear approximation problem -e.g. if

$$X = C[a,b], \quad \| x \| = \max_{t \in [a,b]} |x(t)| \text{ and } V = P_n,$$

the set of polynomials of degree n (or less).

4) We consider a problem in optimal control theory, for which the following question could be the starting point (c.f. LUENBERGER [53, p. 228]): a rocket shall reach a certain minimal altitude using a minimum of fuel, whereby the thrust is limited for technical reasons. More explicitly we have the following problem:

On $M := \{u \in L^2[t_o,t_1] : |u(t)| \leq u_{max} \text{ a.e. on } [t_o,t_1],\ x(t_1) \geq c$

$$\text{with } \dot{x}(t) = A(t)x(t) + b(t)u(t), x(t_o) = x_o\}$$

minimize $f(u) := \frac{1}{2} \int_{t_o}^{t_1} u^2(t)dt$. Here $u_{max} > 0$, $c \in \mathbb{R}^n$ and $A \in C([t_o,t_1],\mathbb{R}^{n\times n})$, $b \in C([t_o,t_1],\mathbb{R}^n)$ are given. If one denotes by $\Phi(\cdot)$ the fundamental system belonging to the differential equation $\dot{x} = A(t)x$ and normalized by $\Phi(t_o) = I$, then for fixed $u \in L^2[t_o,t_1]$ the solution of the initial value problem

$$\dot{x} = A(t)x + b(t)u(t), \quad x(t_o) = x_o$$

is given by

$$x(t) = \Phi(t)x_o + \Phi(t) \int_{t_o}^{t} \Phi^{-1}(s)b(s)u(s)ds.$$

In order to bring this problem into the form of the general convex program (P) we set

$$X := L^2[t_o,t_1], \quad C := \{u \in L^2[t_o,t_1] : |u(t)| \le u_{max} \text{ a.e.}\},$$

$$Y := \mathbb{R}^n, \quad K := \{z \in \mathbb{R}^n : z \ge 0\}$$

and define $g : L^2[t_o,t_1] \to \mathbb{R}^n$ by

$$g(u) := c - \Phi(t_1)x_o - \Phi(t_1) \int_{t_o}^{t_1} \Phi^{-1}(s)b(s)u(s)ds.$$

5) Let $S \subset \mathbb{R}^n$ be compact and

$$f(x) := \max_{y \in S} |x-y|.$$

$f : \mathbb{R}^n \to R$ is convex and the unconstrained convex optimization problem of minimizing $f(x)$ on $M := \mathbb{R}^n$ has a solution $\bar{x}$ (for with respect to an arbitrary $x^o \in \mathbb{R}^n$ the level set

$$W_o := \{x \in \mathbb{R}^n : f(x) \le f(x^o)\}$$

is compact, since it is closed and contained in $B[x^o;2f(x^o)]$) and $B[\bar{x};f(\bar{x})]$ is obviously a circumsphere for S, i.e. a sphere with minimal radius containing S. But S has exactly one circumsphere (proof?).

6) A firm has the objective of optimizing its production plan for the manufacture of a certain product over a certain time interval [O,T]. One assumes the minimum demand for this product over the given time interval is known. Surplus production must

be stored; storage costs are proportional to the stock held. Suppose production costs are proportional to the rate of production. Let

$x(t)$ = stock held at time t

$r(t)$ = rate of production at time t

$d(t)$ = demand at time t.

Then $\dot{x}(t) = r(t) - d(t)$, $x(0) = x_0$ = given initial stock. We must find a production plan given by the function $r(\cdot)$ such that the conditions

$0 \leq r(t)$ (possibly also $r(t) \leq r_{max}$ = max. rate of production)

$$0 \leq x(t) = x_0 + \int_0^t (r(s)-d(s))ds$$

are fulfilled on $[0,T]$ and the objective function

$$f(r) := \int_0^T \{c(r(t)) + hx(t)\}dt$$

$$(= hTx_0 + \int_0^T \{c(r(t)) + h(T-t)(r(t)-d(t))\}dt$$

after partial integration)

is minimal. Here $c(r)$ designates the production cost rate for the production level r, $h > 0$ is a constant and storage costs are given by

$$h \int_0^T x(t)dt.$$

If $c : \mathbb{R} \to \mathbb{R}$ is convex, then this continuous production planning problem is a convex optimization problem. This example can be found in LUENBERGER [53, p. 4 and p. 237] and for special c very extensively treated in ARROW-KARLIN [2].

7) The following optimization problem can be regarded as a (greatly simplified) model for the control of air pollution (details in e.g. KRABS [45, p. 13], KIRSCH-WARTH-WERNER [42,

p. 133] and GUSTAFSON-KORTANEK [30]):

$$\text{Minimize } f : \mathbb{R}^n \to \mathbb{R} \text{ defined by } f(x) := \sum_{j=1}^{n} c_j x_j$$

($c_j > 0$, $j = 1,\ldots,n$ given) subject to

$$0 \le x_j \le 1 \quad (j=1,\ldots,n)$$

$$\sum_{j=1}^{n} x_j u_j(t) \ge v(t) \text{ for all } t \in B.$$

Here $B \subset \mathbb{R}^N$ is compact, u_j, $v \in C(B)$. This is a so-called semiinfinite linear program: the space X in which the solution is sought is finite dimensional, but the set of feasible solutions

$$M := \{x \in \mathbb{R}^n : 0 \le x_j \le 1 \ (j=1,\ldots,n), \sum_{j=1}^{n} x_j u_j(t) \ge v(t) \text{ for all } t \in B\}$$

is in general described by infinitely many side conditions. If however one defines $g : \mathbb{R}^n \to \mathbb{R}$ by

$$g(x) := \max_{t \in B} \{v(t) - \sum_{j=1}^{n} x_j u_j(t)\},$$

then one has an equivalent program

$$\text{Minimize } f(x) := \sum_{j=1}^{n} c_j x_j \text{ subject to}$$

$$x \in C := \{x \in \mathbb{R}^n : 0 \le x \le e\},\ g(x) \le 0$$

where $e = (1,\ldots,1)^T \in \mathbb{R}^n$).

These examples have perhaps also made it clear that it is reasonable to distinguish between the <u>explicit restrictions</u> $x \in C$ (in general simple side conditions, e.g. nonnegativity conditions) and <u>implicit restrictions</u> $g(x) \in -K$.

4.2 Definition and motivation of the dual program. The weak duality theorem

In this section we want to set up a dual program to the program

(P) Minimize $f(x)$ on $M := \{x \in X : x \in C,\ g(x) \in -K\}$,

for which in the next section we shall generally assume that

(A) i) X is a linear space and $f : X \to \mathbb{R}$ is convex

ii) $C \subset X$ is nonempty and convex

iii) $(Y, \|\ \|)$ is a normed linear space, $K \subset Y$ a convex cone and $g : X \to Y$ K-convex.

In this section however we can do without any convexity assumptions. Entirely in analogy to the geometric motivation of the dual linear program in 2.1 we define for (P) the set

$$\Lambda := \{(g(x)+z, f(x)+r) \in Y \times \mathbb{R} : x \in C,\ z \in K,\ r \geq 0\}.$$

Under the assumptions (A) i) - iii) Λ is convex (proof?). Furthermore

1. (P) is feasible, i.e. $M \neq \emptyset \iff \exists\ \beta \in \mathbb{R}$ with $(0,\beta) \in \Lambda$.
2. If (P) is feasible, then

$$\inf\ (P) := \inf\{f(x) : x \in M\} = \inf\{\beta : (0,\beta) \in \Lambda\}.$$

Exactly as in 2.1 the primal problem (P) consists in finding the smallest possible intersection of Λ with the $\mathbb{R}$-axis . Formulated verbally the dual problem consists in determining among all nonvertical closed hyperplanes in $Y \times \mathbb{R}$ which contain Λ in their nonnegative halfspace the one whose intersection with the $\mathbb{R}$-axis is as large as possible.

The nonvertical closed hyperplanes in $Y \times \mathbb{R}$ can be represented by

$$H(y^*,\alpha) := \{(y,r) \in Y \times \mathbb{R} : r + \langle y^*,y\rangle = \alpha\},$$

where $(y^*,\alpha) \in Y^* \times \mathbb{R}$. The corresponding nonnegative halfspace is

$$H^+(y^*,\alpha) := \{(y,r) \in Y \times \mathbb{R} : r + \langle y^*,y\rangle \geq \alpha\}.$$

Obviously

$$\Lambda \subset H^+(y^*,\alpha) \iff f(x) + \langle y^*,g(x)+z\rangle \geq \alpha \text{ for all } x \in C,\ z \in K.$$

Thus as a first version of a dual program we get

$(\tilde{D})$ Maximize α on $\tilde{N} := \{(y^*,\alpha) \in Y^* \times \mathbb{R} : f(x) + \langle y^*,g(x)+z\rangle \geq \alpha$ for all $x \in C,\ z \in K\}$.

If $K \subset Y$ is a cone then from $(y^*,\alpha) \in \tilde{N}$ it follows that $\langle y^*,z\rangle \geq 0$ for all $z \in K$. This suggests the following definition:

<u>4.2.1 Definition:</u> Let $(X,\|\ \|)$ be a normed linear space and $A \subset X$. Then

$$A^+ := \{x^* \in X^* : \langle x^*,a\rangle \geq 0 \text{ for all } a \in A\}$$

is the <u>dual cone</u> of A.

The designations in the literature are not standardized. Instead of dual cone one often speaks of the polar cone or conjugate cone. A^+ is obviously always a convex closed cone in $(X^*,\|\ \|)$, where the norm in X^* is naturally given by

$$\|x^*\| = \sup_{x \neq 0} \frac{|\langle x^*,x\rangle|}{\|x\|}$$

<u>Example:</u> Let

$$K := \{z \in \mathbb{R}^m : z_i = 0\ (i=1,\ldots,m_1),\ z_i \geq 0\ (i=m_1+1,\ldots,m)\}.$$

Then $K^+ = \{y \in \mathbb{R}^m : y_i \geq 0\ (i=m_1+1,\ldots,m)\}$.

If $K \subset Y$ is a cone, then $(\tilde{D})$ is equivalent to

$$\text{(D)} \quad \text{Maximize } \varphi(y^*) := \inf_{x \in C} (f(x) + \langle y^*, g(x) \rangle) \text{ on}$$

$$N := \{y^* \in Y^* : y^* \in K^+, \varphi(y^*) > -\infty\}$$

We call (D) the program dual to (P). However, we should not forget that this is only justified when $K \subset Y$ is a cone. If this is not the case, then we have $(\tilde{D})$ or any equivalent problem as dual program.

Examples: 1) We wish to establish a connection to § 2 and thus consider the general linear program

$$\text{Minimize } c^T x = \sum_{j=1}^{n} c_j x_j \text{ subject to}$$

$$(Ax)_i = \sum_{j=1}^{n} a_{ij} x_j \left\{ \begin{array}{l} = b_i \text{ for } i = 1, \ldots, m_1 \\ \\ \geq b_i \text{ for } i = m_1+1, \ldots, m \end{array} \right.$$

$$x_j \geq 0 \text{ for } j = 1, \ldots, n_1.$$

If we set

$$X = \mathbb{R}^n, \; f(x) = c^T x, \; C = \{x \in \mathbb{R}^n : x_j \geq 0 \; (j=1,\ldots,n_1)\}.$$

$$Y = \mathbb{R}^m, \; g(x) = b - Ax \text{ and}$$

$$K = \{z \in \mathbb{R}^m : z_i = 0 \; (i=1,\ldots,m_1), \; z_i \geq 0 \; (i=m_1+1,\ldots,m)\}$$

then this linear program is of the form

$$\text{(P)} \qquad \text{Minimize } f(x) \text{ on } M := \{x \in X : x \in C, \; g(x) \in -K\}.$$

Since we can identify $(\mathbb{R}^m)^*$ with $\mathbb{R}^m$ the set of feasible solutions of the dual program is given by

$$N = \{y \in \mathbb{R}^m : y_i \geq 0 \; (i=m_1+1,\ldots,m),$$

$$\varphi(y) = \inf_{x \in C} (c^T x + y^T (b - Ax)) > -\infty\}.$$

But $\varphi(y) = b^Ty + \inf_{x\in C} (c-A^Ty)^Tx > -\infty$ precisely when

$$(c-A^Ty)_j = c_j - \sum_{i=1}^{m} a_{ij}y_i \left\} \begin{array}{l} \geq 0 \text{ for } j = 1,\ldots,n_1 \\ \\ = 0 \text{ for } j = n_1+1,\ldots,n, \end{array}\right.$$

and in this case $\varphi(y) = b^Ty$. Thus the dual program is

$$\text{Maximize } b^Ty = \sum_{i=1}^{m} b_iy_i \text{ subject to}$$

$$(A^Ty)_j = \sum_{i=1}^{m} a_{ij}y_i \left\} \begin{array}{l} \leq c_j \text{ for } j = 1,\ldots,n_1 \\ \\ = c_j \text{ for } j = n_1+1,\ldots,n. \end{array}\right.$$

$$y_i \geq 0 \text{ for } i = m_1+1,\ldots,m.$$

This agrees exactly with what we got in 2.1.

2) We shall deal extensively with quadratic programs in 4.4.

3) Let $(X, \|\ \ \|)$ be a normed linear space and $f(x) := \| x-z \|$ for fixed $z \in X$. We wish to consider linear and convex approximation problems and set up the dual program.

a) Let $V \subset X$ be a linear subspace. Suppose given the linear approximation problem of minimizing $f(x)$ on V. As a linear subspace V is also a cone. We can set $Y = X$, $K = V$, $C = X$ and $g(x) = x$. Then

$$K^+ = V^+ = \{x^* \in X^* : \langle x^*,v\rangle = 0 \text{ for all } v \in V\}.$$

For given $x^* \in X^*$ we have

$$\varphi(x^*) = \inf_{x\in X} (\| x-z \| + \langle x^*,x\rangle) > -\infty \iff \| x^* \| \leq 1$$

(proof?) and in this case

$$\varphi(x^*) = \inf_{x\in X} (\| x-z \| + \langle x^*,x-z\rangle) + \langle x^*,z\rangle = \langle x^*,z\rangle.$$

Thus the program dual to the linear approximation problem is

$$\text{Maximize } \varphi(x^*) := \langle x^*, z\rangle \text{ on } N := \{x^* \in X^* : x^* \in V^+, \|x^*\| \le 1\}.$$

b) Let $C \subset X$ be convex. Suppose given the convex approximation problem of minimizing $f(x)$ on C resp.

$$\text{Minimize } f(x) \text{ on } M := \{x \in X : g(x) := x \in -(-C)\}.$$

Since $K := -C$ need not be a cone, we at first have as dual program

$$\text{Maximize } \alpha \text{ on } \tilde{N} := \{(x^*,\alpha) \in X^* \times \mathbb{R} : \|x-z\| + \langle x^*, x-c\rangle \ge \alpha$$

$$\text{for all } x \in X,\ c \in C\}.$$

Obviously $(x^*,\alpha) \in \tilde{N}$ precisely when

$$\sup_{c\in C} \langle x^*, c\rangle < +\infty,\quad \|x^*\| \le 1 \text{ and}$$

$$\varphi(x^*) := \langle x^*, z\rangle - \sup_{c\in C} \langle x^*, c\rangle \ge \alpha.$$

Thus as dual program we get

$$\text{Maximize } \varphi(x^*) := \langle x^*, z\rangle - \sup_{c\in C} \langle x^*, c\rangle \text{ on}$$

$$N := \{x^* \in X^* : \sup_{c\in C} \langle x^*, c\rangle < +\infty,\ \|x^*\| \le 1\}.$$

If $C = V$ is a linear subspace then we obviously have the same result as in a).

The main result of this paragraph is the following weak duality theorem. Geometrically it says that the intersection α of the $\mathbb{R}$-axis and a hyperplane $H(y^*,\alpha) = \{(y,r) \in Y \times \mathbb{R} : r + \langle y^*, y\rangle = \alpha\}$ containing

$$\Lambda = \{(g(x)+z, f(x)+r) \in Y \times \mathbb{R} : x \in C,\ z \in K,\ r \ge 0\}$$

in its nonnegative halfspace is not larger than any intersection of Λ with the $\mathbb{R}$-axis .

<u>4.2.2 Theorem</u> (Weak Duality Theorem): Suppose given the programs

(P) Minimize $f(x)$ on $M = \{x \in X : x \in C,\ g(x) \in -K\}$

and

(D) Maximize $\varphi(y^*) = \inf\limits_{x \in C} (f(x) + \langle y^*, g(x)\rangle)$ on

$$N = \{y^* \in Y^* : y^* \in K^+,\ \varphi(y^*) > -\infty\}.$$

Suppose further that (P) and (D) are feasible, i.e. $M \neq \emptyset$ and $N \neq \emptyset$. If $x \in M$ and $y^* \in N$, then $\varphi(y^*) \leq f(x)$. If $\varphi(y^*) = f(x)$, then x is a solution of (P) and y^* is a solution of (D).

<u>Proof</u>: $\varphi(y^*) \leq f(x) + \langle y^*, g(x)\rangle \leq f(x)$, since $g(x) \in -K$ and $y^* \in K^+$. The rest is obvious.

<u>Remarks</u>: 1. If in Theorem 4.2.2 one replaces (D) by

$(\tilde{D})$ Maximize $\tilde{\varphi}(y^*, \alpha) := \alpha$ on

$$\tilde{N} := \{(y^*, \alpha) \in Y^* \times \mathbb{R} : f(x) + \langle y^*, g(x)+z\rangle \geq \alpha \quad \text{for all } x \in C,\ z \in K\},$$

then the claim of the weak duality theorem naturally still holds mutis mutandem. I.e. if $x \in M$, $(y^*, \alpha) \in \tilde{N}$, then

$$\tilde{\varphi}(y^*, \alpha) = \alpha \leq f(x) + \langle y^*, g(x) + (-g(x))\rangle = f(x).$$

2. We did not need any convexity assumptions for the proof of the weak duality theorem.

3. Just as in 2.1 one can deduce from the weak duality theorem:

i) $\sup (D) \leq \inf (P)$

ii) (D) feasible $\Rightarrow$ inf (P) $> -\infty$

iii) (P) feasible $\Rightarrow$ sup (D) $< +\infty$.

A corresponding statement holds for $(\tilde{D})$ in place of (D) because of the first remark.

4. If sup (D) < inf (P), then one speaks of a duality gap. The geometric interpretation of the dual program makes it seem plausible that in nonconvex problems (resp. on nonconvex sets Λ) there will in general be a duality gap. However even in convex - in fact even in linear and finite dimensional - programs a duality gap can occur as the following example due to FAN [21] shows:

Let $X := Y := \mathbb{R}^3$ and

$$C := K := \{z \in \mathbb{R}^3 : z_1 \geq 0,\ z_2 \geq 0,\ z_1 z_2 \geq z_3^2\}.$$

Then K is a closed convex cone in $\mathbb{R}^3$. We let $f(x) := x_3$ and $g(x) := (0,-1-x_3,-x_1)^T$. For the program

$$\text{(P)} \qquad \text{Minimize } f(x) \text{ on } M = \{x \in \mathbb{R}^3 : x \in C,\ g(x) \in -K\}$$

we have inf (P) = min (P) = 0, since

$$M = \{x \in \mathbb{R}^3 : x_1 = x_3 = 0,\ x_2 \geq 0\}.$$

Now we want to set up the program dual to (P). The cone dual to K is

$$K^+ = \{y \in \mathbb{R}^3 : y_1 z_1 + y_2 z_2 + y_3 z_3 \geq 0 \text{ for all } z \in K\}$$

$$= \{y \in \mathbb{R}^3 : y_1 z_1 + y_2 z_2 \pm y_3 \sqrt{z_1 z_2} \geq 0 \text{ for all } z_1, z_2 \geq 0\}$$

$$= \{y \in \mathbb{R}^3 : y_1 \geq 0,\ y_2 \geq 0,\ 4 y_1 y_2 \geq y_3^2\}.$$

As objective function of the dual program one has

$$\varphi(y) = \inf_{x\in C} (f(x)+y^T g(x))$$

$$= \inf_{x\in C} (x_3-y_2(1+x_3)-y_3x_1)$$

$$= - y_2 + \inf_{x\in C} ((1-y_2)x_3-y_3x_1).$$

Thus $\varphi(y) > -\infty$ if and only if $y_2 = 1$, $y_3 \le 0$, and in this case $\varphi(y) = - y_2$. The program dual to (P) is thus

(D) Maximize $\varphi(y) = - y_2$ on

$$N = \{y \in \mathbb{R}^3 : y_1 \ge 0,\ y_2 = 1,\ y_3 \le 0,\ 4y_1 \ge y_3^2\}.$$

Hence sup (D) = max (D) = - 1; a duality gap occurs.

Using the weak duality theorem one can obtain a lower bound for inf (P) by evaluating the objective function of the dual program in a point feasible for the dual problem. In principle this lower bound can be made arbitrarily exact if no duality gap is present.

Example: We already considered above the linear approximation problem of minimizing $f(x) = \| x-z \|$ on a linear subspace V of the normed linear space $(X, \| \ \|)$ and found as dual program

(D) Maximize $\langle x^*, z\rangle$ on

$$N := \{x^* \in X^* : \langle x^*, v\rangle = 0 \text{ for all } v \in V, \| x^* \| \le 1\}.$$

Now we look at a special case. Let

$$X = C[a,b],\ \| x \| = \max_{t\in[a,b]} |x(t)| \text{ and } V \subset C[a,b]$$

be an (n+1)-dimensional linear subspace fulfilling the HAAR condition: every not identically vanishing $v \in V$ has at most n zeros in [a,b]. One then calls V an <u>(n+1)-dimensional HAAR system</u> on [a,b]. An example of such a system is $V = \mathcal{P}_n$. We wish to construct nontrivial elements in N. To this end choose

$$T := \{t_o,\dots,t_{n+1}\} \subset [a,b] \text{ with } a \le t_o < t_1 < \dots < t_{n+1} \le b.$$

Suppose $V = \text{span}\ \{v_o,\dots,v_n\}$. Then up to a factor ± 1 there exists exactly one $q = (q_o,\dots,q_{n+1})^T$ with

$$\sum_{k=0}^{n+1} v_j(t_k)q_k = 0 \quad (j=0,\dots,n), \quad \sum_{k=0}^{n+1} |q_k| = 1.$$

For:

$$\sum_{k=0}^{n+1} v_j(t_k)q_k = 0 \iff \sum_{k=0}^{n} v_j(t_k)q_k = - v_j(t_{n+1})q_{n+1}$$

$$(j=0,\dots,n) \qquad\qquad (j=0,..,n)$$

$$\iff \begin{pmatrix} q_o \\ \cdot \\ \cdot \\ \cdot \\ q_n \end{pmatrix} = -q_{n+1}(v_j(t_k))^{-1} \begin{pmatrix} v_o(t_{n+1}) \\ \cdot \\ \cdot \\ \cdot \\ v_n(t_{n+1}) \end{pmatrix}$$

The matrix $(v_j(t_k))_{0\le j,k\le n}$ here is nonsingular, since V satisfies the HAAR condition. Thus q_{n+1} and hence also q is determined up to a factor ± 1 by the additional requirement

$$\sum_{k=0}^{n+1} |q_k| = 1.$$

Now define

$$\rho := \sum_{k=0}^{n+1} z(t_k)q_k \text{ and } x^* : C[a,b] \to \mathbb{R} \text{ by}$$

$$<x^*,x> := \text{sign}\ (\rho) \sum_{k=0}^{n+1} x(t_k)q_k.$$

Then we have: 1. x^* is linear. From

$$|<x^*,x>| \le \sum_{k=0}^{n+1} |x(t_k)|\,|q_k| \le \sum_{k=0}^{n+1} |q_k|\,\|x\| = \|x\|$$

for all $x \in C[a,b]$ it follows $x^* \in (C[a,b])^*$ and $\|x^*\| \le 1$.

2. $\langle x^*, v\rangle = 0$ for all $v \in V$.

Thus x^* is feasible for the dual problem and the weak duality theorem gives

$$\langle x^*, z\rangle = \left|\sum_{k=0}^{n+1} z(t_k) q_k\right| \leq \inf_{x\in V} \| x-z \| .$$

We wish to illustrate this construction with a numerical example.

Let $[a,b] = [\frac{1}{2},1]$, $z(t) \equiv 1$, $V = \text{span}\,\{1/\sqrt{t}, \sqrt{t}\}$. The corresponding approximation problem was considered already in 1.3. V is obviously a 2-dimensional HAAR system. We choose $T = \{\frac{1}{2}, t_1, 1\}$ with $\frac{1}{2} < t_1 < 1$. $q = (q_o, q_1, q_2)^T$ is to be determined by

$$\sqrt{2}\, q_o + \frac{1}{\sqrt{t_1}} q_1 + q_2 = 0 \qquad\qquad |q_1| + |q_2| + |q_3| = 1$$

$$\frac{1}{\sqrt{2}} q_o + \sqrt{t_1}\, q_1 + q_2 = 0.$$

From these equations one obtains

$$\begin{pmatrix} q_o \\ q_1 \\ q_2 \end{pmatrix} = \pm \frac{1}{\sqrt{2}-1+\sqrt{t_1}+(2-\sqrt{2})t_1} \begin{pmatrix} \sqrt{2}\,(1-t_1) \\ -\sqrt{t_1} \\ 2t_1 - 1 \end{pmatrix}.$$

Then it follows that

$$\left|\sum_{k=0}^{2} q_k\right| = \frac{\sqrt{t_1}-(\sqrt{2}-1)(1+\sqrt{2}t_1)}{\sqrt{t_1}+(\sqrt{2}-1)(1+\sqrt{2}t_1)}$$

$$\leq \inf_{A,B\in\mathbb{R}} \max_{t\in[\frac{1}{2},1]} \left|\frac{A+Bt-\sqrt{t}}{\sqrt{t}}\right| .$$

By appropriate choice of $t_1 \in (\frac{1}{2},1)$ one tries to make the lower bound as large as possible. The maximum is achieved for $t_1 = 1/\sqrt{2}$ and one obtains

$$\left(\frac{1-(1/2)^{1/4}}{1+(1/2)^{1/4}}\right)^2 \le \inf_{A,B\in\mathbb{R}} \max_{t\in[\frac{1}{2},1]} \left|\frac{A+Bt-\sqrt{t}}{\sqrt{t}}\right|$$

(One can show that in this case equality holds).

4.3 Strong duality, KUHN-TUCKER saddle point theorem

In this section we consider convex programs (this time the convexity is really decisive) of the form

(P) Minimize $f(x)$ on $M := \{x \in X : x \in C,\ g(x) \in -K\}$

and the dual problem

(D) Maximize $\varphi(y^*) := \inf_{x\in C} (f(x)+\langle y^*, g(x)\rangle)$ on

$N := \{y^* \in Y^* : y^* \in K^+,\ \varphi(y^*) > -\infty\}$

Under the assumptions

(A) i) X is a linear space and $f : X \to \mathbb{R}$ is convex

ii) $C \subset X$ is nonempty and convex

iii) $(Y, \|\ \ \|)$ is a normed linear space, $K \subset Y$ a convex cone and $g : X \to Y$ is K-convex

and a suitable further condition we shall show that no duality gap can occur.

4.3.1 Theorem: Suppose given the programs (P) and (D) defined above. Suppose further that (A) i) - iii) are fulfilled. Moreover suppose

$$\Lambda := \{(g(x)+z, f(x)+r) \in Y \times \mathbb{R} : x \in C,\ z \in K,\ r \ge 0\}$$

is closed. Then we have

i) (P) is feasible, $\inf (P) > -\infty \iff$ (D) is feasible, $\sup (D) < +\infty$.
In both cases (P) has a solution and

$$- \infty < \sup (D) = \min (P) < + \infty.$$

ii) (P) not feasible, (D) feasible $\Rightarrow \sup (D) = + \infty$.

iii) (P) feasible, (D) not feasible $\Rightarrow \inf (P) = - \infty$.

Proof: i) 1) Suppose (P) is feasible and $\inf (P) > - \infty$. Then $(0,\inf(P)) \in \Lambda$, i.e. (P) has a solution. For there exists a sequence $\{x_k\} \subset M$ with $f(x_k) \to \inf (P)$. Because Λ is closed and $\{(0,f(x_k))\} \subset \Lambda$ it follows that $(0,\inf(P)) \in \Lambda$. Thus we can write min (P) instead of inf (P). Now choose an $\alpha < \min(P)$ arbitrarily. Then $(0,\alpha) \notin \Lambda$. Since Λ is convex and closed we can apply the strict separation theorem 3.2.5. It says that we can strictly separate Λ and $\{(0,\alpha)\}$ by a closed hyperplane in $Y \times \mathbb{R}$. Thus there exist $(y^*,\lambda_o) \in Y^* \times \mathbb{R} \setminus \{(0,0)\}$, $\gamma \in \mathbb{R}$ with

$$\lambda_o \alpha < \gamma < \lambda_o (f(x)+r) + \langle y^*,g(x)+z \rangle$$

for all $x \in C$, $z \in K$, $r \geq 0$. Since $(0,\min(P)) \in \Lambda$ we have in particular

$$\lambda_o \alpha < \gamma < \lambda_o \min (P)$$

Since $\alpha < \min (P)$ we must have $\lambda_o > 0$, i.e. the hyperplane separating $\{(0,\alpha)\}$ and Λ in $Y \times \mathbb{R}$ is nonvertical. Without restriction we may assume $\lambda_o = 1$. Then

$$\alpha < \gamma < f(x) + \langle y^*,g(x)+z \rangle \text{ for all } x \in C,\ z \in K.$$

Since K is a cone, $y^* \in K^+$ and moreover

$$\alpha < \varphi(y^*) = \inf_{x \in C} (f(x)+\langle y^*,g(x) \rangle).$$

Thus, $y^* \in N$ is feasible for the dual program and $\alpha < \varphi(y^*) \leq \sup (D)$. Since $\alpha < \min (P)$ was arbitrary we have $\min (P) \leq \sup (D)$. The weak duality theorem 4.2.2 implies $\sup (D) = \min (P)$.

2) Suppose (D) is feasible and $\sup (D) < + \infty$. We show that then $(0,\sup(D)) \in \Lambda$ ($\Rightarrow$ (P) is feasible and $\inf (P) > - \infty$). For assume this were not the case. Then $\{(0,\sup(D))\}$ and Λ can

be strictly separated by a closed hyperplane in $Y \times \mathbb{R}$. There exist $(y_o^*, \lambda_o) \in Y^* \times \mathbb{R} \setminus \{(0,0)\}$, $\gamma \in \mathbb{R}$ with

$$\lambda_o \sup (D) < \gamma < \lambda_o(f(x)+r) + \langle y_o^*, g(x)+z\rangle$$

$$\text{for all } x \in C,\ z \in K,\ r \geq 0.$$

Then $y_o^* \in K^+$, $\lambda_o \geq 0$ and $\lambda_o \sup (D) < \gamma < \lambda_o f(x) + \langle y_o^*, g(x)\rangle$ for all $x \in C$.

We distinguish two cases:

If $\lambda_o > 0$, then $y^* := \frac{1}{\lambda_o} y_o^* \in N$ and $\sup (D) < \varphi(y^*)$, a contradiction.

If on the other hand $\lambda_o = 0$, then $0 < \gamma < \langle y_o^*, g(x)\rangle$ for all $x \in C$. By assumption (D) is feasible, so there exists a $y^* \in N$. For all $t \geq 0$ we have $y^* + ty_o^* \in N$ and $\varphi(y^*+ty_o^*) \geq \varphi(y^*) + t\gamma$. Letting t tend to $+\infty$ we get a contradiction to $\sup (D) < +\infty$.

ii), iii) follow immediately from i).

<u>Remarks</u>: 1. If in Theorem 4.3.1 one gives up the assumptions (A) i) - iii) and instead assumes that

$$\Lambda = \{(g(x)+z, f(x)+r) \in Y \times \mathbb{R} : x \in C,\ z \in K,\ r \geq 0\}$$

is convex, then the statement of the theorem still holds if one replaces (D) by

$(\tilde{D})$ Maximize $\tilde{\varphi}(y^*, \alpha) := \alpha$ on

$$\tilde{N} := \{(y^*, \alpha) \in Y^* \times \mathbb{R} : \alpha \leq f(x) + \langle y^*, g(x)+z\rangle$$

$$\text{for all } x \in C,\ z \in K\}$$

(proof?)

2. In particular under the assumptions of the last theorem one has: If (P) and (D) are feasible, then (P) has a solution and $\min (P) = \sup (D)$.

3. If in Theorem 4.3.1 instead of assuming Λ is closed one

assumes

$$\text{cl}\ (L \cap \Lambda) = L \cap \text{cl}\ (\Lambda)$$

with $L := \{(0,\beta) \in Y \times \mathbb{R} : \beta \in \mathbb{R}\}$ (such programs are called normal in VAN SLYKE-WETS [75] and KRABS [45]), then the assertions of Theorem 4.3.1 still hold; one must do without the solvability of (P) and write inf (P) instead of min (P).

Examples: 1) Suppose given a finite dimensional linear program in normal form

$$\text{Minimize } c^T x \text{ on } M := \{x \in \mathbb{R}^n : Ax = b,\ x \geq 0\}.$$

With $X = \mathbb{R}^n$, $f(x) = c^T x$, $C = \{x \in \mathbb{R}^n : x \geq 0\}$, $Y = \mathbb{R}^m$, $K = \{0\}$ and $g(x) = b - Ax$ the assumptions (A) i) - iii) are fulfilled. Furthermore

$$\Lambda = \{(g(x)+z, f(x)+r) \in Y \times \mathbb{R} : x \in C,\ z \in K,\ r \geq 0\}$$

$$= (b,0) + \{(-Ax, c^T x+r) : x \geq 0,\ r \geq 0\}$$

is closed since

$$\{(-Ax, c^T x+r) : x \geq 0,\ r \geq 0\} = \{\begin{pmatrix} -A & 0 \\ c^T & 1 \end{pmatrix}\begin{pmatrix} x \\ r \end{pmatrix} : \begin{pmatrix} x \\ r \end{pmatrix} \geq 0\}$$

is a finitely generated cone (c.f. Definition 2.2.5) and as such closed by Theorem 2.2.6. From Theorem 4.3.1 we thus recover the statement of the strong duality theorem in linear optimization. One should not however think this approach is simpler than the direct approach employed in 2.3. There we needed only the FARKAS lemma, here the FARKAS lemma (to prove that finitely generated cones are closed) and the strict separation theorem (for the proof of 4.3.1). A similar situation prevails in the duality theory of quadratic programing as we shall see in the next section. There too one can obtain the strong duality theorem by appealing to 4.3.1, but a direct proof is simpler.

2) We wish to return to Example 4) in 4.1 and shall use the notation introduced there. We want to prove that

$$\Lambda = \{(g(u)+z, f(u)+r) \in \mathbb{R}^n \times \mathbb{R} : u \in C,\ z \in K,\ r \geq 0\}$$

is closed. In order not to have to make an excessively long exposition here we use without proof a statement that will follow easily later from more general results in 6.2:

$$C = \{u \in L^2[t_o, t_1] : |u(t)| \leq u_{max} \text{ a.e. on } [t_o, t_1]\}$$

is <u>weakly sequentially compact</u> in $L^2[t_o,t_1]$. That is: for every sequence $\{u_i\} \subset C$ there exists a subsequence $\{u_{i_k}\} \subset \{u_i\}$ and a $u \in C$ with

$$\int_{t_o}^{t_1} u_{i_k}(t)v(t)dt \to \int_{t_o}^{t_1} u(t)v(t)dt \text{ for all } v \in L^2[t_o,t_1].$$

(One says: $\{u_{i_k}\}$ converges weakly to u and writes $u_{i_k} \rightharpoonup u$).

Now the proof that Λ is closed is not hard. Let

$$\{(g(u_i) + z_i, f(u_i) + r_i)\} \subset \Lambda$$

be a sequence converging to $(y,q) \in \mathbb{R}^n \times \mathbb{R}$. Let $\{u_{i_k}\} \subset \{u_i\}$ be a subsequence converging weakly to $u \in C$. Then obviously

$$g(u_{i_k}) = c - \Phi(t_1)x_o - \Phi(t_1) \int_{t_o}^{t_1} \Phi^{-1}(t)b(t)u_{i_k}(t)dt$$

$$\to c - \Phi(t_1)x_o - \Phi(t_1) \int_{t_o}^{t_1} \Phi^{-1}(t)b(t)u(t)dt = g(u).$$

Because $g(u_{i_k}) + z_{i_k} \to y$ and the nonnegative orthant K in $\mathbb{R}^n$ is closed we have $z_{i_k} \to z := y - g(u) \in K$. It remains to show that $f(u) \leq q$, for then $(y,q) = (g(u)+z, f(x)+(q-f(u)) \in \Lambda$. But

$$f(u) + \int_{t_o}^{t_1} (u_{i_k}(t)-u(t))u(t)dt \le f(u) + \frac{1}{2} \int_{t_o}^{t_1} (u_{i_k}^2(t)-u^2(t))dt$$

$$= f(u_{i_k})$$

$$\le f(u_{i_k}) + r_{i_k}.$$

The left hand side converges to $f(u)$, the right to q and hence $f(u) \le q$. Thus we have proved that Λ is closed and may use 4.3.1.

Let us now set up the dual problem. The objective function is

$$\varphi(y) = \inf_{u \in C} (f(u)+y^T g(u))$$

$$= y^T(c-\Phi(t_1)x_o) - \frac{1}{2} \int_{t_o}^{t_1} \eta^2(t)dt$$

$$+ \inf_{u \in C} \frac{1}{2} \int_{t_o}^{t_1} (u(t)-\eta(t))^2 dt,$$

where we have abbreviated $\eta(t) := y^T\Phi(t_1)\Phi^{-1}(t)b(t)$. Obviously

$$\inf_{u \in C} \frac{1}{2} \int_{t_o}^{t_1} (u(t)-\eta(t))^2 dt = \frac{1}{2} \int_{t_o}^{t_1} (v(t)-\eta(t))^2 dt \text{ with}$$

$$v(t) = \begin{cases} -u_{max} & \text{for } \eta(t) < -u_{max} \\ \eta(t) & \text{for } |\eta(t)| \le u_{max} \\ u_{max} & \text{for } \eta(t) > u_{max} \end{cases}$$

Thus we have determined the (rather complicated) objective function of the dual program. It is to be maximized on

$$N = \{y \in \mathbb{R}^n : y \ge 0\}.$$

3) Let V be a finite dimensional linear subspace of a normed linear space $(X, \| \ \|)$. As in Example 3) a) of 4.2 we consider the linear approximation problem of minimizing $f(x) = \| x-z \|$ for fixed $z \in X$ subject to $g(x) := x \in V$ $(= -V)$. As dual prog-

ram we obtained in 4.2:

$$\text{Maximize } \langle x^*, z\rangle \text{ on } N = \{x^* \in X^* : x^* \in V^+, \| x^* \| \leq 1\}.$$

We wish to show that the hypotheses of 4.3.1 are satisfied. Thus we have to show that

$$\Lambda = \{(x+v, \| x-z \| + r) : x \in X,\ v \in V,\ r \geq 0\}$$

is closed. To this end let $\{(x_i+v_i, \| x_i-z \| + r_i)\} \subset \Lambda$ be a sequence converging to the point $(y,q) \in X \times \mathbb{R}$. $\{x_i\}$ is bounded since $\| x_i-z \| + r_i \to q$ and $\{v_i\}$ is bounded since $x_i + v_i \to y$. Since V is finite dimensional we can choose a subsequence $\{v_{i_k}\} \subset \{v_i\}$ converging say to $v \in V$. Then $\{x_{i_k}\}$ converges to $x := y - v$ and we have

$$(y,q) = (x+v, \| x-z \| + (q - \| x-z \|)) \in \Lambda.$$

Hence no duality gap occurs and the finite dimensional linear approximation problem has a solution.

Theorem 4.3.1 says nothing about the solvability of the dual problem. About this we prove:

<u>4.3.2 Theorem</u>: Suppose given the programs (P) and (D) defined above. Suppose the assumptions (A) i) - iii) are fulfilled. For

$$\Lambda := \{(g(x)+z, f(x)+r) \in Y \times \mathbb{R} : x \in C,\ z \in K,\ r \geq 0\}$$

suppose moreover $\operatorname{int}(\Lambda) \cap \{0\} \times \mathbb{R} \neq \emptyset$. Then (P) is feasible and we have:

i) If $\inf(P) > -\infty$, then (D) has a solution $\bar{y}^*$ and $\max(D) = \inf(P)$ holds.

ii) If (P) has a solution $\bar{x}$, then $\langle \bar{y}^*, g(\bar{x})\rangle = 0$.

<u>Proof</u>: i) Suppose $\inf(P) > -\infty$. Then $(0, \inf(P)) \notin \operatorname{int}(\Lambda)$. The separation theorem 3.2.4 implies the existence of

$(\bar{y}^*, \lambda_o) \in Y^* \times \mathbb{R} \setminus \{(0,0)\}$ with

1. $\lambda_o \inf (P) \leq \lambda_o(f(x)+r) + \langle \bar{y}^*, g(x)+z \rangle$

$$\text{for all } x \in C,\ z \in K,\ r \geq 0$$

2. $\lambda_o \inf (P) < \lambda_o q + \langle \bar{y}^*, y \rangle$ for all $(y,q) \in \operatorname{int} (\Lambda)$.

From 1. it follows $\bar{y}^* \in K^+$ and $\lambda_o \geq 0$. Since $\operatorname{int} (\Lambda) \cap \{0\} \times \mathbb{R} \neq \emptyset$ there is a $q \in \mathbb{R}$ with $(0,q) \in \operatorname{int} (\Lambda)$, so from 2. it follows $\lambda_o > 0$. Without loss of generality we may assume $\lambda_o = 1$ and thus

$$\inf (P) \leq f(x) + \langle \bar{y}^*, g(x) \rangle \text{ for all } x \in C.$$

Thus $\bar{y}^*$ is feasible for the dual problem and

$$\inf (P) \leq \inf_{x \in C} (f(x) + \langle \bar{y}^*, g(x) \rangle) = \varphi(\bar{y}^*) \leq \sup (D)$$

By the weak duality theorem $\sup (D) \leq \inf (P)$. Therefore $\bar{y}^*$ is a solution of (D) and $\max (D) = \inf (P)$.

ii) If $\bar{x}$ is a solution of (P), then

$$f(\bar{x}) = \inf (P) \leq f(\bar{x}) + \langle \bar{y}^*, g(\bar{x}) \rangle \leq f(\bar{x}),$$

and hence $\langle \bar{y}^*, g(\bar{x}) \rangle = 0$.

Remarks: 1. Parallel to the first remark following Theorem 4.3.1 we have: If instead of (A) i) - iii) one has that

$$\Lambda = \{(g(x)+z, f(x)+r) \in Y \times \mathbb{R} : x \in C,\ z \in K,\ r \geq 0\}$$

is convex (if one no longer demands for example that $K \subset Y$ be a cone), then the assertion of Theorem 4.3.2 i) still holds if one replaces (D) by

$(\tilde{D})$ Maximize $\tilde{\varphi}(y^*, \alpha) = \alpha$ on

$$\tilde{N} = \{(y^*,\alpha) \in Y^* \times \mathbb{R} : \alpha \le f(x) + \langle y^*, g(x)+z\rangle \text{ for all } x \in C,\ z \in K\}$$

(proof?). If moreover $\bar{x}$ is a solution of (P) and $(\bar{y}^*, \inf(P))$ a solution of $(\tilde{D})$, then $0 \le \langle \bar{y}^*, g(\bar{x})+z\rangle$ for all $z \in K$.

2. The condition $\operatorname{int}(\Lambda) \cap \{0\} \times \mathbb{R} \neq \emptyset$ in Theorem 4.3.2 is fulfilled if e.g. the so-called SLATER constraint qualification holds, i.e. if

a) $\operatorname{int}(K) \neq \emptyset$

b) There exists an $\hat{x} \in C$ with $g(\hat{x}) \in -\operatorname{int}(K)$.

For then we have

$$\{(g(x)+z, f(x)+r) \in Y \times \mathbb{R} : x \in C, z \in \operatorname{int}(K), r > 0\} \subset \operatorname{int}(\Lambda)$$

and thus in particular

$$(g(\hat{x}) + (-g(\hat{x})), f(\hat{x})+r) = (0, f(\hat{x})+r) \in \operatorname{int}(\Lambda) \cap \{0\} \times \mathbb{R}$$

for all $r > 0$.

Any supplemental condition that guarantees that a hyperplane separating $\{(0,\inf(P))\}$ and Λ in $Y \times \mathbb{R}$ is nonvertical is called a constraint qualification. For finite dimensional problems there are several other constraint qualifications in addition to that of SLATER. The reader is referred to MANGASARIAN [57, p. 78] and BAZARAA-SHETTY [3, p. 160].

Examples: 1) As in Example 3) b) in 4.2 we consider once again the convex approximation problem

$$\text{(P)} \qquad \text{Minimize } f(x) := \| x-z \| \text{ on } C.$$

Here $(X, \| \ \|)$ shall be a normed linear space, $C \subset X$ nonempty and convex and $z \in X$. We wrote (P) in the form

$$\text{(P)} \quad \text{Minimize } f(x) = \| x-z \| \text{ on } M := \{x \in X : g(x) := x \in -(-C)\}$$

and obtained as dual program

(D) Maximize $\varphi(x^*) := \langle x^*, z\rangle - \sup_{c\in C} \langle x^*, c\rangle$ on

$$N := \{x^* \in X^* : \sup_{c\in C} \langle x^*, c\rangle < +\infty, \| x^* \| \leq 1\}.$$

Since the set

$$\Lambda = \{(x-c, \| x-z \| + r) \in X \times \mathbb{R} : x \in X,\ c \in C,\ r \geq 0\}$$

belonging to (P) is convex and

$$\{(x-c, \| x-z \| + r) : x \in X,\ c \in C,\ r > 0\} \subset \text{int}\,(\Lambda),$$

i.e. $\text{int}\,(\Lambda) \cap \{0\} \times \mathbb{R} \neq \emptyset$, it follows from Theorem 4.3.2 -or more precisely from the following remark- that (D) is solvable and max (D) = inf (P). If moreover $\bar{x}$ is a solution of (P) and $\bar{x}^*$ a solution of (D), then $0 \leq (\bar{x}^*, \bar{x}-c)$ for all $c \in C$ resp.

$$\sup_{c\in C} \langle \bar{x}^*, c\rangle = \langle \bar{x}^*, \bar{x}\rangle \text{ and thus}$$

$$\langle \bar{x}^*, z-\bar{x}\rangle = \varphi(\bar{x}^*) = \max\,(D) = \inf\,(P) = \| z-\bar{x} \|.$$

2) In Example 6) in 4.1 we presented the continuous production planning problem. It was:

$$\text{Minimize } f(r) := hTx_o + \int_0^T \{c(r(t)) + h(T-t)(r(t)-d(t))\}dt$$

$$\text{on the set } M := \{r \in C_{pc}[0,T] : 0 \leq r(t) \text{ and}$$

$$-x_o + \int_0^t (d(s)-r(s))ds \leq 0 \text{ on } [0,T]\}.$$

Here $C_{pc}[0,T]$ is the set of piecewise continuous functions on $[0,T]$, $d \in C_{pc}[0,T]$ a given nonnegative function (demand function) and $x_o \geq 0$ (initial stock); moreover $c : \mathbb{R} \to \mathbb{R}$ shall be convex, e.g. $c(r) = \frac{1}{2} r^2$. If one defines $X := C_{pc}[0,T]$,

$$C := \{r \in X : r(t) \geq 0 \text{ on } [0,T]\},\ Y := C[0,T]$$

$$g : X \to Y \text{ by } g(r) := -x_o + \int_0^t (d(s)-r(s))ds \quad \text{and}$$

$$K := \{x \in C[0,T] : x(t) \geq 0 \text{ on } [0,T]\},$$

then the hypotheses (A) i) - iii) are satisfied. Theorem 4.3.2 is applicable if $x_o > 0$, for then the SLATER constraint qualification holds. Thus if inf (P) > $-\infty$, then the dual problem is solvable and no duality gap occurs. But what is the dual problem? To answer this question we need the (topological) dual space of C[0,T] and K^+. Information on this topic can be found e.g. in LUENBERGER [53, p. 113]. We do not wish to go into this topic and leave a discussion of the dual program as an exercise.

We can now prove a strong duality theorem for finite dimensional convex programs with inequalities and (affine linear) equations as restrictions.

<u>4.3.3 Theorem</u>: Consider the problem

(P) Minimize f(x) on

$$M := \{x \in \mathbb{R}^n : x \in C,\ g_1(x) \in -K_1,\ g_2(x) = 0\}.$$

Assume that:

i) $f : \mathbb{R}^n \to \mathbb{R}$ is convex.

ii) $C \subset \mathbb{R}^n$ is convex.

iii) $K_1 \subset \mathbb{R}^m$ is a convex cone with int $(K_1) \neq \emptyset$, $g_1 : \mathbb{R}^n \to \mathbb{R}^m$ is K_1-convex and $g_2 : \mathbb{R}^n \to \mathbb{R}^k$ is affine linear, i.e. $g_2(x) = G_2 x + h_2$ with $G_2 \in \mathbb{R}^{k\times n}$ and $h_2 \in \mathbb{R}^k$.

Moreover assume that the following constraint qualification holds:

a) There exists an $\hat{x} \in C$ with $g_1(\hat{x}) \in -\text{int}\ (K_1)$, $g_2(\hat{x}) = 0$.

b) $0 \in \text{int}\ g_2(C)$.

Then we have: If inf (P) > $-\infty$, then the dual program

(D) Maximize $\varphi(u_1,u_2) := \inf_{x\in C} (f(x)+u_1^T g_1(x)+u_2^T g_2(x))$

on $N := \{(u_1,u_2) \in \mathbb{R}^m \times \mathbb{R}^k : u_1 \in K_1^+, \varphi(u_1,u_2) > -\infty\}$

(K_1^+ is of course $\{u_1 \in \mathbb{R}^m : u_1^T z_1 \geq 0$ for all $z_1 \in K_1\}$) has a solution $(\bar{u}_1,\bar{u}_2)$ and

$$\inf (P) = \varphi(\bar{u}_1,\bar{u}_2) = \max (D).$$

If (P) has a solution $\bar{x}$, then $\bar{u}_1^T g_1(\bar{x}) = 0$.

Proof: It is possible to apply Theorem 4.3.2. We prefer to give a direct proof and define

$$\Lambda_+ := \{(g_1(x)+z_1, g_2(x), f(x)+r) \in \mathbb{R}^m \times \mathbb{R}^k \times \mathbb{R} \\ : x \in C,\ z_1 \in K_1, r > 0\}.$$

Then Λ_+ is nonempty and convex and $(0,0,\inf(P)) \notin \Lambda_+$. By theorem 3.1.9 two disjoint convex sets -here $\{(0,0,\inf(P))\}$ and Λ_+- in a finite dimensional space -here $\mathbb{R}^m \times \mathbb{R}^k \times \mathbb{R}$- can be separated by a hyperplane. Thus there exist $(\bar{u}_1,\bar{u}_2,\bar{u}_0) \in \mathbb{R}^m \times \mathbb{R}^k \times \mathbb{R}\setminus\{(0,0,0)\}$ with

$$\bar{u}_0 \inf (P) \leq \bar{u}_0(f(x)+r) + \bar{u}_1^T(g_1(x)+z_1) + \bar{u}_2^T g_2(x)$$
$$\text{for all } x \in C, z_1 \in K_1, r > 0.$$

Obviously $\bar{u}_0 \geq 0$, $\bar{u}_1 \in K_1^+$ must hold. Suppose we had $\bar{u}_0 = 0$, i.e.

$$0 \leq \bar{u}_1^T(g_1(x)+z_1) + \bar{u}_2^T g_2(x) \text{ for all } x \in C,\ z_1 \in K_1.$$

From hypothesis a) one gets $\bar{u}_1 = 0$, and then from b) that also $\bar{u}_2 = 0$ -altogether a contradiction to $(\bar{u}_1,\bar{u}_2,\bar{u}_0) \neq (0,0,0)$. Thus without loss of generality we may assume $\bar{u}_0 = 1$. It follows $\inf (P) \leq \varphi(\bar{u}_1,\bar{u}_2)$ and $(\bar{u}_1,\bar{u}_2) \in N$. If we furthermore take into account the weak duality theorem, i.e.

$$\varphi(\bar{u}_1,\bar{u}_2) \leq \sup (D) \leq \inf (P),$$

then we get $\inf\ (P) = \varphi(\bar{u}_1,\bar{u}_2) = \max\ (D)$. The rest follows as in Theorem 4.3.2.

Theorem 4.3.2 can also be formulated as a saddle point theorem for the LAGRANGE-functional $L : X \times Y^* \to \mathbb{R}$,

$$L(x,y^*) := f(x) + \langle y^*,g(x)\rangle.$$

We define:

4.3.4 Definition: Suppose given the convex program

(P) Minimize $f(x)$ on $M := \{x \in X : x \in C,\ g(x) \in -K\}$

under the hypotheses (A) i) - iii). Then $(\bar{x},\bar{y}^*) \in C \times K^+$ is called a saddle point of the LAGRANGE-functional

$$L(x,y^*) = f(x) + \langle y^*,g(x)\rangle$$

if $L(\bar{x},y^*) \le L(\bar{x},\bar{y}^*) \le L(x,\bar{y}^*)$ for all $x \in C$, $y^* \in K^+$

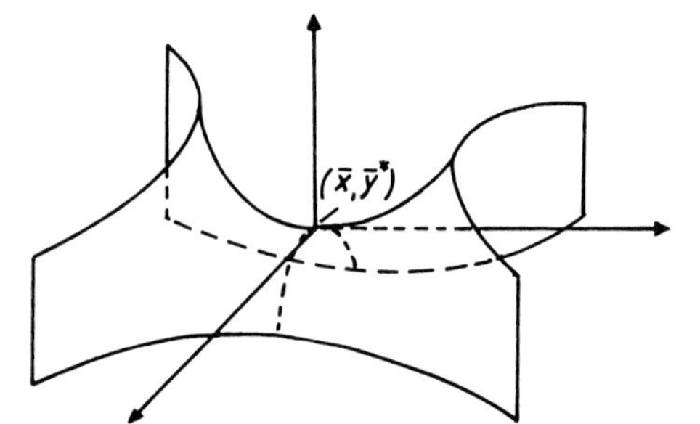

Thus if $(\bar{x},\bar{y}^*)$ is a saddle point of L, then $L(x,\cdot)$ on K^+ assumes a maximum at $\bar{y}^*$ and $L(\cdot,\bar{y}^*)$ on C assumes a minimum at $\bar{x}$.

4.3.5 Theorem (KUHN-TUCKER Saddle Point Theorem): Suppose given the convex program

(P) Minimize $f(x)$ on $M := \{x \in X : x \in C,\ g(x) \in -K\}$

under the hypotheses (A) i) - iii). Then we have

i) Suppose $\operatorname{int}(\Lambda) \cap \{0\} \times \mathbb{R} \neq \emptyset$ with

$$\Lambda := \{(g(x)+z, f(x)+r) \in Y \times \mathbb{R} : x \in C,\ z \in K,\ r \ge 0\}$$

(e.g. the SLATER constraint qualification fulfilled: there exists an $\hat{x} \in C$ with $g(\hat{x}) \in -\operatorname{int}(K)$). If $\bar{x} \in M$ is a solution of (P), then there exists a $\bar{y}^* \in K^+$ such that $(\bar{x},\bar{y}^*)$ is a saddle point of the LAGRANGE-functional L.

ii) If K is closed and $(\bar{x},\bar{y}^*) \in C \times K^+$ is a saddle point of the LAGRANGE-functional L, then $\bar{x}$ is a solution of (P) (in particular $\bar{x} \in M$ resp. $g(\bar{x}) \in -K$).

Proof: i) Let $\bar{y}^* \in K^+$ be a solution of the dual program, which by Theorem 4.3.2 exists. For this solution we have $\langle\bar{y}^*,g(\bar{x})\rangle = 0$. Thus

$$\begin{aligned} L(\bar{x},y^*) &= f(\bar{x}) + \langle y^*,g(\bar{x})\rangle \\ &\le f(\bar{x}) \quad \text{for } y^* \in K^+ \text{ (since } g(\bar{x}) \in -K) \\ &= L(\bar{x},\bar{y}^*) \\ &\le f(x) + \langle\bar{y}^*,g(x)\rangle = L(x,\bar{y}^*) \quad \text{for all } x \in C. \end{aligned}$$

ii) Suppose $-g(\bar{x}) \notin K$. Since by hypothesis K is closed, the strict separation theorem 3.2.5 gives the existence of a $y_0^* \in Y^*$, $\gamma \in \mathbb{R}$ with

$$-\langle y_0^*,g(\bar{x})\rangle < \gamma < \langle y_0^*,z\rangle \text{ for all } z \in K.$$

Then $y_0^* \in K^+$ and $\langle y_0^*,g(\bar{x})\rangle > 0$. If one sets $y^* := \bar{y}^* + y_0^*$, then $L(\bar{x},y_0^*) > L(\bar{x},\bar{y}^*)$, a contradiction. Thus $\bar{x} \in M$. Since $(\bar{x},\bar{y}^*)$ is a saddle point of L and $g(\bar{x}) \in -K$ we have $\langle\bar{y}^*,g(\bar{x})\rangle = 0$. Thus for arbitrary $x \in M$

$$f(\bar{x}) = L(\bar{x},\bar{y}^*) \le L(x,\bar{y}^*) = f(x) + \langle\bar{y}^*,g(x)\rangle \le f(x),$$

i.e. $\bar{x}$ is a solution of (P).

For finite dimensional convex programs with inequality restrictions we give still another version of the last two theorems which uses the concept of the subdifferential.

Reminder: If $f \in \operatorname{Conv}(\mathbb{R}^n)$, i.e. $f : \mathbb{R}^n \to \mathbb{R}$ convex, then

$$\partial f(x) := \{l \in \mathbb{R}^n : l^T(z-x) \leq f(z) - f(x) \text{ for all } z \in \mathbb{R}^n\}$$

is called the subdifferential of f at $x \in \mathbb{R}^n$ (c.f. Definition 3.3.4).

In the following lemma we collect several properties of the subdifferential.

4.3.6 Lemma: i) Suppose $f \in \text{Conv}(\mathbb{R}^n)$. Then $\partial f(x)$ is non-empty, convex and compact for every $x \in \mathbb{R}^n$.

ii) If $f_1, f_2 \in \text{Conv}(\mathbb{R}^n)$, then $\partial(f_1+f_2)(x) = \partial f_1(x) + \partial f_2(x)$ for every $x \in \mathbb{R}^n$ (MOREAU-ROCKAFELLAR).

Proof: i) By Theorem 3.3.6 we have $\partial f(x) \neq \emptyset$. The convexity and closedness of $\partial f(x)$ follows directly from the definition of the subdifferential. The boundedness of $\partial f(x)$ follows easily from the continuity of f (c.f. Corollary 3.3.9): if $l \in \partial f(x)$ then

$$l^T h \leq f(x+h) - f(x) \leq \max_{|h| \leq 1} f(x+h) - f(x) =: C(x)$$

for arbitrary $h \in B[0;1]$ and thus $|l| \leq C(x)$ for every $l \in \partial f(x)$.

ii) Since the inclusion $\partial f_1(x) + \partial f_2(x) \subset \partial(f_1+f_2)(x)$ is trivial we only need to show that every $l \in \partial(f_1+f_2)(x)$ can be represented in the form $l = l_1 + l_2$ with $l_1 \in \partial f_1(x)$, $l_2 \in \partial f_2(x)$.

Let us define

$$C_1 := \{(y, f_1(x+y) - f_1(x) + r_1) \in \mathbb{R}^n \times \mathbb{R} : y \in \mathbb{R}^n, r_1 \geq 0\}$$

$$C_2 := \{(z, l^T z - f_2(x+z) + f_2(x) - r_2) \in \mathbb{R}^n \times \mathbb{R} : z \in \mathbb{R}^n, r_2 > 0\}.$$

Then C_1, C_2 are nonempty and convex and since $l \in \partial(f_1+f_2)(x)$ we have $C_1 \cap C_2 = \emptyset$. Hence C_1 and C_2 can be separated by a hyperplane in $\mathbb{R}^n \times \mathbb{R}$ (Theorem 3.1.9). Thus there exists

$$(l_1, \lambda_o) \in \mathbb{R}^n \times \mathbb{R} \setminus \{(0,0)\} \text{ with}$$

$$l_1^T y - \lambda_o(f_1(x+y)-f_1(x)+r_1)$$

$$\leq l_1^T z - \lambda_o(l^T z-f_2(x+z)+f_2(x)-r_2)$$

for all $y,z \in \mathbb{R}^n$, $r_1 \geq 0$, $r_2 > 0$.

Evidently $\lambda_o \geq 0$. Since $(l_1,\lambda_o) \neq (0,0)$ we have $\lambda_o > 0$ and without loss of generality we may assume $\lambda_o = 1$:

$$l_1^T y - (f_1(x+y)-f_1(x)) \leq l_1^T z - (l^T z-f_2(x+z)+f_2(x))$$

for all $y,z \in \mathbb{R}^n$. If one sets $z = 0$ on the right side of this inequality one gets $l_1 \in \partial f_1(x)$; if one sets $y = 0$ on the left side then it follows $l - l_1 \in \partial f_2(x)$, which proves that

$$\partial(f_1+f_2)(x) \subset \partial f_1(x) + \partial f_2(x).$$

The following is the consequence of the KUHN-TUCKER saddle point theorem mentioned earlier.

<u>4.3.7 Theorem</u>: Suppose $f,g_i \in \text{Conv}(\mathbb{R}^n)$ $(i=1,\ldots,m)$ and suppose given the convex program

(P) Minimize $f(x)$ on $M := \{x \in \mathbb{R}^n : g_i(x) \leq 0 \ (i=1,..,m)\}$.

Then we have: i) If the SLATER constraint qualification is fulfilled, i.e. there exists an $\hat{x} \in \mathbb{R}^n$ with $g_i(\hat{x}) < 0$ $(i=1,\ldots,m)$ and if $\overline{x} \in M$ is a solution of (P), then there exists a $\overline{y} = (\overline{y}_i) \in \mathbb{R}^m$ with $\overline{y} \geq 0$, $\overline{y}_i g_i(\overline{x}) = 0$ $(i=1,\ldots,m)$ and

$$0 \in \partial f(\overline{x}) + \overline{y}_1 \partial g_1(\overline{x}) + \ldots + \overline{y}_m \partial g_m(\overline{x})$$

ii) Suppose $\overline{x} \in M$. If there exists a $\overline{y} = (\overline{y}_i) \in \mathbb{R}^m$ with $\overline{y} \geq 0$, $\overline{y}_i g_i(\overline{x}) = 0$ $(i=1,\ldots,m)$ and

$$0 \in \partial f(\overline{x}) + \overline{y}_1 \partial g_1(\overline{x}) + \ldots + \overline{y}_m \partial g(\overline{x}),$$

then $\overline{x}$ is a solution of (P).

<u>Proof</u>: i) The KUHN-TUCKER saddle point theorem 4.3.5 is applicable and supplies the existence of a $\bar{y} = (\bar{y}_i) \in \mathbb{R}^m$ with $\bar{y} \geq 0$ and

$$f(\bar{x}) + \sum_{i=1}^{m} y_i g_i(\bar{x}) \leq f(\bar{x}) + \sum_{i=1}^{m} \bar{y}_i g_i(\bar{x})$$

$$\leq f(x) + \sum_{i=1}^{m} \bar{y}_i g_i(x)$$

for all $x \in \mathbb{R}^n$, $y \in \mathbb{R}^m$ with $y \geq 0$. Then $\bar{y}_i g_i(\bar{x}) = 0$ and

$$h := f + \sum_{i=1}^{m} \bar{y}_i g_i \in \text{Conv}\,(\mathbb{R}^n)$$

assumes its minimum on $\mathbb{R}^n$ at $\bar{x}$. By definition of the subgradient $0 \in \partial h(\bar{x})$. By Lemma 4.3.6 ii) and since $\partial(\lambda g)(x) = \lambda \partial g(x)$ for $\lambda \geq 0$ we have

$$0 \in \partial h(\bar{x}) = \partial f(\bar{x}) + \partial(\bar{y}_1 g_1)(\bar{x}) + \ldots + \partial(\bar{y}_m g_m)(\bar{x})$$

$$= \partial f(\bar{x}) + \bar{y}_1 \partial g_1(\bar{x}) + \ldots + \bar{y}_m \partial g_m(\bar{x}).$$

ii) Suppose $0 = l + \sum_{i=1}^{m} \bar{y}_i l_i$ with $l \in \partial f(\bar{x})$, $l_i \in \partial g_i(\bar{x})$. Then for arbitrary $x \in M$ we have:

$$f(x) - f(\bar{x}) \geq l^T(x-\bar{x}) = - \sum_{i=1}^{m} \bar{y}_i l_i^T(x-\bar{x})$$

$$\geq - \sum_{i=1}^{m} \bar{y}_i (g_i(x) - g_i(\bar{x}))$$

$$\geq 0$$

since $\bar{y}_i g_i(\bar{x}) = 0$ and $\bar{y}_i g_i(x) \leq 0$ $(i=1,\ldots,m)$. Thus $\bar{x}$ is a solution of (P).

<u>Remark</u>: If in the last theorem the g_i are affine linear one can skip the hypothesis of the constraint qualification (proof?).

To close this section we give three applications each of which will be formulated as a theorem, since each is interesting in its own right. In order to compute the subdifferentials that occur we need a consequence of the Theorem of CARATHEODORY.

4.3.8 Lemma: If $A \subset \mathbb{R}^n$ is compact then the convex hull co (A) of A is also compact.

Proof: Let $S := \{\lambda = (\lambda_1,\dots,\lambda_{n+1}) : \lambda_i \geq 0 \ (i=1,\dots,n+1), \sum_{i=1}^{n+1} \lambda_i = 1\}$.

Then $S \subset \mathbb{R}^{n+1}$ is compact. Define a map

$$\Psi : S \times \underbrace{A \times A \times \dots \times A}_{n\,+\,1\text{ factors}} \to \text{co}\,(A)$$

by $\Psi((\lambda_1,\dots,\lambda_{n+1}),a^1,\dots,a^{n+1}) = \sum_{i=1}^{n+1} \lambda_i a^i$. Then by the Theorem of CARATHEODORY (2.2.7) we have $\Psi(S \times A \times \dots \times A) = \text{co}\,(A)$, so co (A) is the continuous image of a compact set $S \times A \times \dots \times A$ and is thus compact itself.

As a first consequence we prove as in IOFFE-TICHOMIROV [37, p. 352ff] an old result related to a problem posed by J.J. SYLVESTER 1857: "It is required to find the least circle which shall contain a given system of points in a plane".

4.3.9 Theorem (JUNG, 1901): Suppose $S \subset \mathbb{R}^n$ is bounded. Further let

$$\overline{R} := \inf\{R : \exists x \text{ with } S \subset B[x;R]\}$$

be the circumradius and

$$\overline{D} := \sup_{y,z \in S} |y-z|$$

be the diameter of S. Then we have

$$\overline{R} \leq \overline{D}\left(\frac{n}{2(n+1)}\right)^{1/2} .$$

Proof: Obviously we can without restriction assume that S is compact and contains more than one point. Define $f : \mathbb{R}^n \to \mathbb{R}$ by

$$f(x) := \max_{y \in S} |x-y| .$$

Then $f \in \text{Conv}(\mathbb{R}^n)$. As we already saw earlier (c.f. Example 5) in 4.1) there exists an $\bar{x} \in \mathbb{R}^n$ with $f(\bar{x}) \leq f(x)$ for all $x \in \mathbb{R}^n$. $B[\bar{x};\bar{R}]$ with $\bar{R} = f(\bar{x})$ is the circumsphere for S. Furthermore $0 \in \partial f(\bar{x})$. In order to make full use of this necessary and sufficient optimality condition we need a representation of the subdifferential.

1) We have

$$\partial f(x) = \text{co}\, \{ \frac{x-u}{|x-u|} : u \in S(x) \},$$

where $S(x) := \{u \in S : |x-u| = f(x)\}$. (Since S contains more than one point $f(x) > 0$ for all $x \in \mathbb{R}^n$.)

i) Suppose $u \in S(x)$ and $z \in \mathbb{R}^n$. Then

$$\frac{(x-u)}{|x-u|}^T (z-x) = \frac{(x-u)}{|x-u|}^T (z-u) - f(x)$$

$$\leq f(z) - f(x) \Rightarrow \frac{x-u}{|x-u|} \in \partial f(x).$$

Since $\partial f(x)$ is convex we have

$$\text{co}\, \{ \frac{x-u}{|x-u|} : u \in S(x) \} \subset \partial f(x).$$

ii) Since S(x) as a closed subset of the compact set S is itself compact, it follows that

$$A := \{ \frac{x-u}{|x-u|} : u \in S(x) \}$$

is compact. By Lemma 4.3.8 co (A) is also compact. Suppose there were an $l \in \partial f(x) \setminus \text{co}\,(A)$. From the strict separation theorem 3.2.5 follows the existence of a $y \in \mathbb{R}^n$ and $\varepsilon > 0$ with

$$l^T y \geq \frac{(x-u)}{|x-u|}^T y + \varepsilon \quad \text{for all } u \in S(x).$$

Let $t_k \searrow 0$. Choose $\{u_k\} \subset S$ with

$$f(x+t_k y) = |x+t_k y-u_k| .$$

Since S is compact we can without loss of generality assume that u_k converges to a u_o. Obviously $u_o \in S(x)$. Since $l \in \partial f(x)$, we have

$$\frac{f(x+t_k y)-f(x)}{t_k} \geq l^T y \geq \frac{(x-u_o)^T}{|x-u_o|} y + \varepsilon .$$

Because

$$\lim_{k\to\infty} \frac{|x+t_k y-u_o|-|x-u_o|}{t_k} = \frac{(x-u_o)^T}{|x-y_o|} y$$

there exists a $k_o \in \mathbb{N}$ with

$$\frac{|x+t_{k_o} y-u_o|-|x-u_o|}{t_{k_o}} \leq \frac{(x-u_o)^T}{|x-u|} y + \frac{\varepsilon}{2} .$$

Thus for every $k \geq k_o$:

$$\begin{aligned}
\frac{|x+t_{k_o} y-u_o|-|x-u_o|}{t_{k_o}} + \frac{\varepsilon}{2} &\leq \frac{(x-u_o)^T}{|x-u_o|} y + \varepsilon \\
&\leq \frac{f(x+t_k y)-f(x)}{t_k} \\
&\leq \frac{|x+t_k y-u_k|-|x-u_k|}{t_k} \\
&\leq \frac{|x+t_{k_o} y-u_k|-|x-u_k|}{t_{k_o}}
\end{aligned}$$

(where the last inequality follows from the fact that

$$\varphi(t) := \frac{|x+ty-u_k|-|x-u_k|}{t}$$

is nondecreasing on $(0,\infty)$; c.f. proof of Theorem 3.3.2) and hence

$$t_{k_o} \frac{\varepsilon}{2} \leq |x-u_o| - |x-u_k| + |x+t_{k_o} y-u_k| - |x+t_{k_o} y-u_o|$$

Letting k tend to ∞ we get a contradiction, so 1) is proved.

2) Since $0 \in \partial f(\bar{x})$ and because of 1) and the Theorem of CARATHEODORY there exist an $m \leq n+1$, $u^i \in S(\bar{x})$, $\lambda_i > 0$ $(i=1,\ldots,m)$ with $\sum_{i=1}^{m} \lambda_i = 1$ and $0 = \sum_{i=1}^{m} \lambda_i(\bar{x}-u^i)$. Since

$$f(\bar{x}) = |\bar{x}-u^i| = \bar{R} \text{ for } i = 1,\ldots,m$$

we have for arbitrary $j \in \{1,\ldots,m\}$

$$\sum_{i=1}^{m} \lambda_i |u^i-u^j|^2 = \sum_{i=1}^{m} \lambda_i |(\bar{x}-u^j)-(\bar{x}-u^i)|^2$$

$$= 2\bar{R}^2 - 2\left(\sum_{i=1}^{m} \lambda_i(\bar{x}-u^i)\right)^T(\bar{x}-u^j) = 2\bar{R}^2$$

Thus $$2\bar{R}^2 = \sum_{i,j=1}^{m} \lambda_i\lambda_j|u^i-u^j|^2$$

$$\leq \bar{D}^2 \sum_{\substack{i,j=1\\ i\neq j}}^{m} \lambda_i\lambda_j = \bar{D}^2\left(\left(\sum_{i=1}^{m}\lambda_i\right)^2 - \sum_{i=1}^{m}\lambda_i^2\right)$$

$$\leq \bar{D}^2\left(1-\frac{1}{m}\right) \leq \bar{D}^2\,\frac{n}{n+1}$$

which proves JUNG's inequality

$$\bar{R} \leq \bar{D}\left(\frac{n}{2(n+1)}\right)^{1/2}$$

Remark: If S is a regular simplex, then obviously equality holds in the JUNG inequality.

We return now to the "air pollution problem" in Example 7) in 4.1 and prove:

4.3.10 Theorem: Suppose given the semi-infinite linear program

(P) Minimize $f(x) := \sum_{j=1}^{n} c_j x_j$ on

$$M := \{x \in \mathbb{R}^n : \sum_{j=1}^{n} x_j u_j(t) \geq v(t) \text{ for all } t \in B,\ 0 \leq x_j \leq 1 \quad (j=1,\ldots,n)\}.$$

We assume:

1. $B \subset \mathbb{R}^N$ is compact, u_j, $v \in C(B)$.
2. $c_j > 0$, $u_j(t) \geq 0$ for all $t \in B$, $u_j \not\equiv 0$ $(j=1,\ldots,n)$.
3. $\sum_{j=1}^{n} u_j(t) > v(t)$ for all $t \in B$. Moreover there shall exist a $\hat{t} \in B$ with $v(\hat{t}) > 0$.

Then $\overline{x} \in M$ is a solution of (P) if and only if there exist $m \leq n + 1$, $\lambda_i > 0$ and

$$t_i \in B(\overline{x}) := \{t \in B : v(t) - \sum_{j=1}^{n} \overline{x}_j u_j(t) = 0\}$$

$(i=1,\ldots,m)$ with

$$c_j \begin{Bmatrix} \geq \\ = \\ \leq \end{Bmatrix} \sum_{i=1}^{n} \lambda_i u_j(t_i) \text{ if } \overline{x}_j \begin{Bmatrix} = 0 \\ \in (0,1) \\ = 1 \end{Bmatrix} \qquad (j=1,\ldots,n)$$

<u>Proof</u>: We abbreviate with

$$c = (c_1,\ldots,c_n)^T,\ u(t) = (u_1(t),\ldots,u_n(t))^T,\ e = (1,\ldots,1)^T,$$

$$g(x) := \max_{t \in B} (v(t) - u(t)^T x)$$

and write (P) in the form

(P) Minimize $f(x) = c^T x$ on

$$M := \{x \in \mathbb{R}^n : g(x) \leq 0,\ -x \leq 0,\ x - e \leq 0\}.$$

Since M is compact, (P) has a solution $\overline{x}$. We wish to apply Theorem 4.3.7. The SLATER condition is obviously fulfilled under our hypotheses (let $\hat{x} = (1-\varepsilon)e$ for sufficiently small $\varepsilon > 0$). Thus we get:

$\bar{x} \in M$ is a solution of (P) if and only if there exist $\bar{y}^o, \bar{y}^1 \in \mathbb{R}^n$, $\bar{y}_o \in \mathbb{R}$ with

i) $\bar{y}^o, \bar{y}^1 \geq 0$, $\bar{y}_o \geq 0$

ii) $\bar{y}^{oT}\bar{x} = 0$, $\bar{y}^{1T}(e-\bar{x}) = 0$, $\bar{y}_o g(\bar{x}) = 0$

iii) $0 \in c + \bar{y}_o \partial g(\bar{x}) - \bar{y}^o + \bar{y}^1$

We necessarily have $\bar{y}_o > 0$ and thus $g(\bar{x}) = 0$. For:

$$\bar{y}_o = 0 \underset{iii)}{\Rightarrow} c = \bar{y}^o - \bar{y}^1.$$

Since $0 \notin M$ (by 3.) and $c_j > 0$ $(j=1,\ldots,n)$ we have

$$0 < c^T\bar{x} = \bar{y}^{oT}\bar{x} - \bar{y}^{1T}\bar{x} \underset{ii)}{=} -\bar{y}^{1T}e \underset{i)}{\leq} 0.$$

a contradiction.

Moreover $\partial g(\bar{x}) = -\operatorname{co}\{u(t) \in \mathbb{R}^n : t \in B(\bar{x})\}$ with

$$B(\bar{x}) = \{t \in B : v(t) - u(t)^T\bar{x} = g(\bar{x}) \ (=0)\}.$$

This one shows very much as in the proof of the corresponding statement in the last theorem. The details of the proof are left as an exercise. An application of the Theorem of CARATHEODORY shows that iii) holds if and only if there exist $m \leq n+1$, $t_i \in B(\bar{x})$, $\lambda_i > 0$ $(i=1,\ldots,m)$ with $c - \bar{y}^o + \bar{y}^1 = \sum_{i=1}^{m} \lambda_i u(t_i)$. The assertion then follows from i) and ii).

We conclude this section by proving necessary and sufficient optimality conditions for linear CHEBYSHEV approximation.

4.3.11 Theorem: Suppose $V = \operatorname{span}\{v_1,\ldots,v_n\}$ is an n-dimensional linear subspace of $C(B)$, $B \subset \mathbb{R}^N$ compact, and $z \in C(B) \setminus V$. Suppose given the linear CHEBYSHEV approximation problem

$$\text{(P)} \qquad \text{Minimize } f(x) := \max_{t \in B} |x(t)-z(x)| \text{ on } V.$$

Then we have

i) $\bar{x} \in V$ is a solution of (P) if and only if

$$0 \in \operatorname{co}\{\operatorname{sign}(\bar{x}(t)-z(t)) \begin{pmatrix} v_1(t) \\ \cdot \\ \cdot \\ \cdot \\ v_n(t) \end{pmatrix} : t \in B(\bar{x})\}$$

where $B(\bar{x}) = \{t \in B : |\bar{x}(t)-z(t)| = f(\bar{x})\}$.

ii) Suppose further that $B = [a,b]$ is a compact interval and V an n-dimensional HAAR system on $[a,b]$, i.e. every $v \in V\setminus\{0\}$ has at most $n - 1$ zeros on $[a,b]$. Then $\bar{x} \in V$ is a solution of (P) if and only if there are $n + 1$ points $t_j \in [a,b]$ $(j=1,\ldots,n+1)$ with

α) $$a \leq t_1 < t_2 < \ldots < t_{n+1} \leq b$$

β) $$|\bar{x}(t_j)-z(t_j)| = \max_{t\in[a,b]} |\bar{x}(t)-z(t)|$$

γ) $$\bar{x}(t_j) - z(t_j) = (-1)^{j+1}(\bar{x}(t_1)-z(t_1)) \quad (j=1,\ldots,n+1).$$

<u>Proof</u>: i) By identifying $x = \sum_{i=1}^{n} x_i v_i \in V$ with $(x_i) \in \mathbb{R}^n$ and setting $v(t) = (v_1(t),\ldots,v_n(t))^T$ we can reduce (P) to the unrestricted convex optimization problem

$$\text{Minimize } f(x) := \max_{t\in[a,b]} |v(t)^T x-z(t)| \text{ on } \mathbb{R}^n .$$

$\bar{x} \in \mathbb{R}^n$ is a solution if and only if $0 \in \partial f(\bar{x})$. However

$$\partial f(\bar{x}) = \operatorname{co}\{\operatorname{sign}(v(t)^T\bar{x}-z(t))v(t) : t \in B(\bar{x})\}$$

with

$$B(\bar{x}) := \{t \in B : |v(t)^T\bar{x}-z(t)| = \max_{t\in[a,b]} |v(t)^T\bar{x}-z(t)|\}.$$

For: $\operatorname{co}\{\operatorname{sign}(v(t)^T\bar{x}-z(t))v(t) : t \in B(\bar{x})\} \subset \partial f(\bar{x})$ is again easy. If one assumes there were an $l \in \partial f(\bar{x})$ which does not lie in the convex, compact (Lemma 4.3.8) set

$$\operatorname{co}\{\operatorname{sign}(v(t)^T\bar{x}-z(t))v(t) : t \in B(\bar{x})\}$$

then the strict separation theorem gives the existence of a $y \in \mathbb{R}^n$, $\varepsilon > 0$ with

$$1^T y \geq \text{sign}(v(t)^T \bar{x} - z(t)) v(t)^T y + \varepsilon \text{ for all } t \in B(\bar{x}).$$

For $\{t_k\}$ tending to 0 from above there exists a sequence $\{s_k\} \subset B$ with

$$\max_{t \in B} |v(t)^T (\bar{x} + t_k y) - z(t)| = |v(s_k)^T (\bar{x} + t_k y) - z(s_k)|.$$

Since B is compact one can choose a convergent subsequence of $\{s_k\}$. Without loss of generality we may assume $\{s_k\}$ itself converges: $s_k \to s_o$. Then $s_o \in B(\bar{x})$ and

$$\frac{f(\bar{x}+t_k y) - f(\bar{x})}{t_k} \geq 1^T y \geq \text{sign}(v(s_o)^T \bar{x} - z(s_o)) v(s_o)^T y + \varepsilon$$

Since $z \notin V$ we have $v(s_o)^T \bar{x} - z(s_o) \neq 0$. Thus for all sufficiently large k

$$\begin{aligned} \text{sign}(v(s_o)^T \bar{x} - z(s_o)) &= \text{sign}(v(s_k)^T \bar{x} - z(s_k)) \\ &= \text{sign}(v(s_k)^T (\bar{x} + t_k y) - z(s_k)) \end{aligned}$$

and thus

$$\begin{aligned} \frac{f(\bar{x}+t_k y) - f(\bar{x})}{t_k} &\leq \frac{|v(s_k)^T (\bar{x}+t_k y) - z(s_k)| - |v(s_k)^T \bar{x} - z(s_k)|}{t_k} \\ &= \text{sign}(v(s_o)^T \bar{x} - z(s_o)) v(s_k)^T y \end{aligned}$$

so that

$$\begin{aligned} &\text{sign}(v(s_o)^T \bar{x} - z(s_o)) v(s_o)^T y + \varepsilon \\ &\quad \leq \text{sign}(v(s_o)^T \bar{x} - z(s_o)) v(s_k)^T y \end{aligned}$$

for all sufficiently large k. Letting k tend to ∞ we get the desired contradiction. i) is thus proved.

ii) By i) and the Theorem of CARATHEODORY $\overline{x} \in V$ is a solution of (P) if and only if there exist $m \leq n+1$, $t_j \in B(\overline{x})$, $\lambda_j > 0$ $(j=1,\ldots,m)$ with $\sum_{j=1}^{m} \lambda_j = 1$ and

$$0 = \sum_{j=1}^{m} \lambda_j \, \mathrm{sign}(\overline{x}(t_j)-z(t_j))\, v(t_j).$$

We may suppose the t_j ordered by size:

$$a \leq t_1 < t_2 < \ldots < t_m \leq b.$$

We show:

1) $m = n+1$

2) $q_j := \lambda_j \, \mathrm{sign}(\overline{x}(t_j)-z(t_j))$ alternates in sign, i.e. $q_j q_{j+1} < 0$ $(j=1,\ldots,n)$.

1) Suppose $m < n+1$. Choose pairwise distinct $t_{m+1},\ldots,t_n \in [a,b]$ which are also distinct from $t_1,\ldots,t_m$. Then the system of equations

$$\sum_{j=1}^{n} v_i(t_j)\, y_j = 0 \quad (i=1,\ldots,n)$$

has a nontrivial solution - namely $y = (q_1,\ldots,q_m,0,\ldots,0)^T$. Hence

$$(v_i(t_j))_{1\leq i,j\leq n}$$

is singular, so there exists a $v \in V \setminus \{0\}$ with the n zeros $t_1,\ldots,t_n \in [a,b]$, contradicting the assumption that V should be an n-dimensional HAAR system on [a,b].

2) Since $\sum_{j=1}^{n+1} v_i(t_j) q_j = 0$ $(i=1,\ldots,n)$ we have $\sum_{j=1}^{n+1} u(t_j) q_j = 0$ for all $u \in V$. Let $k \in \{1,\ldots,n\}$ be arbitrary. Define $u \in V$ by the interpolation conditions

$$u(t_j) = \begin{cases} 0 & \text{for } j \in \{1,\ldots,n+1\} \setminus \{k,k+1\} \\ 1 & \text{for } j = k. \end{cases}$$

Since V is an n-dimensional HAAR system on [a,b] u is uniquely determined by these conditions. Furthermore $u(t_{k+1}) > 0$, since u cannot have more than n - 1 zeros. Thus

$$0 = \sum_{j=1}^{n+1} u(t_j)q_j = q_k + u(t_{k+1})q_{k+1}$$

which implies 2). Hence the theorem is proved.

The second part of Theorem 4.3.11 is the Alternation Theorem in CHEBYSHEV approximation. It says that an $\bar{x} \in V$ is the best approximation to a $z \in C[a,b]$ with respect to V if and only if the defect $d(t) := \bar{x}(t) - z(t)$ achieves the maximum of its absolute value in n + 1 successive points t_j in [a,b] and the sign at these points alternates. Precisely this is the basis for the important REMEZ method.

4.4 Quadratic programming

The problem of minimizing a quadratic objective function f, say $f(x) = c^Tx + \frac{1}{2} x^TQx$ with $c \in \mathbb{R}^n$ and symmetric, positive semidefinite $Q \in \mathbb{R}^{n\times n}$, on $\mathbb{R}^n$ subject to finitely many affine linear equality -and inequality- restrictions is called a quadratic program. It is our goal to show that for quadratic programs we can make existence and duality statements that are completely analogous to those for linear programs. Before we begin, though, we give two examples.

Examples: 1) In the method of least squares one wants to minimize the function

$$f(x) := \frac{1}{2} |y-Ax|^2 = \frac{1}{2} x^TA^TAx - (A^Ty)^Tx + \frac{1}{2} y^Ty$$

for given $y \in \mathbb{R}^m$ and $A \in \mathbb{R}^{m\times n}$. If in addition there are linear side conditions, say of the form $-\gamma e \le y - Ax \le \gamma e$ with $e = (1,\ldots,1)^T \in \mathbb{R}^m$ and given $\gamma > 0$ (one could speak of the method of least squares with bounded errors, c.f. KRABS [46]) then one has a quadratic program to solve.

2) The following description of the so-called KRUGER-THIEMER-model is due to PIERCE-SCHUMITZKY [64]:

We consider the control of a simple two-compartment model for drug distribution in the human body. It is assumed that the drug, which is administered orally, is first dissolved into the gastro-intestinal tract, is then absorbed into the so-called apparent volume of distribution (a lumped compartment which accounts for blood, muscle, tissue etc.), and finally is eliminated from the system by the kidneys.

Denote by $x(t)$ and $y(t)$ the amount of drug at time t in the gastro-intestinal tract and apparent volume of distribution, respectively, and by $k_1 > 0$ and $k_2 > 0$ the relevant rate constants. For simplicity, assume that $k_1 \neq k_2$. The dynamical description of this model is then given by

$$(*) \qquad \dot{x} = -k_1 x, \quad \dot{y} = -k_2 + k_1 x.$$

At discrete instants of time $0 = \tau_o < \tau_1 < \dots < \tau_N$ the drug is ingested in amounts $\delta_o, \delta_1, \dots, \delta_N$. This imposes the boundary conditions

$$(**) \qquad \begin{aligned} x(\tau_i+0) &= x(\tau_i-0) + \delta_i \\ y(\tau_i+0) &= y(\tau_i-0) \qquad (i=0,\dots,N) \\ (x(\tau_o-0) &= y(\tau_o-0) = 0). \end{aligned}$$

To achieve a desired therapeutic effect, it is required that the amount of drug in the apparent volume of distribution never goes below a constant level or plateau m, say, during the time interval $[\tau_1, \tau_{N+1}]$, where $\tau_{N+1} > \tau_N$. Thus we have the constraint

$$y(t) \geq m, \quad t \in [\tau_1, \tau_{N+1}].$$

It is also assumed that only nonnegative amounts of drug can be given. Finally, take the biological cost function

$$f(\delta) = \frac{1}{2} |\delta|^2 = \frac{1}{2} \sum_{i=0}^{N} \delta_i^2$$

both to minimize side effects and the cost of the drug. If

$$\begin{pmatrix} \mathbf{x}(\delta;t) \\ y(\delta;t) \end{pmatrix}$$

for given $\delta = (\delta_i) \in \mathbb{R}^{N+1}$ is the solution of (*), (**) then one has the problem

$$\text{Minimize } f(\delta) := \frac{1}{2} |\delta|^2 \text{ subject to}$$

$$\delta \geq 0 \text{ and } y(\delta;t) \geq m \text{ for } t \in [\tau_1, \tau_{N+1}].$$

One easily convinces oneself that

$$y(\delta;t) = \sum_{j=0}^{i} \varphi(t-\tau_j)\delta_j \text{ for } t \in [\tau_i, \tau_{i+1}]$$

with

$$\varphi(t) := \frac{k_1}{k_2-k_1} (e^{-k_1 t} - e^{-k_2 t}).$$

For $\delta \geq 0$ we have

$$\min_{t \in [\tau_i, \tau_{i+1}]} y(\delta;t) = \min(y(\delta;\tau_i), y(\delta;\tau_{i+1})).$$

(For: we may assume that not all δ_j, $j = 0,\dots,i$ vanish. Assume $y(\delta;\cdot)$ takes its minimum on $[\tau_i, \tau_{i+1}]$ at an inner point $\tau \in (\tau_i, \tau_{i+1})$. Then $\dot{y}(\delta;\tau) = 0$, $0 \leq \ddot{y}(\delta;\tau) = k_1 x(\delta;\tau)$ $= - k_1^2 x(\delta;\tau) < 0$, since

$$x(\delta;\tau) = \sum_{j=0}^{i} e^{-k_1(\tau-\tau_j)} \delta_j,$$

a contradiction.) Thus for all $\delta \geq 0$ the restriction $y(\delta;t) \geq m$ for all $t \in [\tau_1, \tau_{N+1}]$ is equivalent to

$$y(\delta;\tau_{i+1}) = \sum_{j=0}^{i} \varphi(\tau_{i+1}-\tau_j)\delta_j \geq m \quad (i=0,\dots,N).$$

If one defines $\Phi = (\Phi_{ij}) \in \mathbb{R}^{(N+1)\times(N+1)}$ by

$$\Phi_{ij} := \begin{cases} \varphi(\tau_{i+1}-\tau_j) & \text{for } i \geq j \\ 0 & \text{for } i < j \end{cases}.$$

and $e = (1,\ldots,1)^T \in \mathbb{R}^{N+1}$, then one has to solve the quadratic program

$$\text{Minimize } f(\delta) = \frac{1}{2}\,\delta^T\delta \text{ subject to}$$

$$\delta \geq 0,\ \Phi\delta \geq me.$$

As in the linear case we begin with a quadratic program in normal form - namely

$$\text{(P)} \qquad \text{Minimize } f(x) := c^Tx + \frac{1}{2}\,x^TQx$$

$$\text{on } M := \{x \in \mathbb{R}^n : Ax = b,\ x \geq 0\}.$$

Here $c \in \mathbb{R}^n$, $Q \in \mathbb{R}^{n\times n}$ is symmetric and positive semidefinite, $A \in \mathbb{R}^{m\times n}$ and $b \in \mathbb{R}^m$.

By setting $X = \mathbb{R}^n$, $C = \{x \in \mathbb{R}^n : x \geq 0\}$, $Y = \mathbb{R}^m$, $K = \{0\}$ and $g(x) = b - Ax$ we can bring (P) into the form of the general convex program investigated in the last section

$$\text{Minimize } f(x) \text{ on } M := \{x \in X : x \in C,\ g(x) \in -K\}.$$

Remark: The general quadratic program

$$\text{Minimize } c_1^Tx_1 + \frac{1}{2}\,x_1^TQ_1x_1 + c_2^Tx_2 + \frac{1}{2}\,x_2^TQ_2x_2 \text{ subject to}$$

$$\begin{array}{l} A_{11}x_1 + A_{12}x_2 = b_1 \\ A_{21}x_1 + A_{22}x_2 \geq b_2 \end{array} \quad ,\ x_1 \geq 0$$

is equivalent to

$$\text{Minimize } \begin{pmatrix} c_1 \\ c_2 \\ -c_2 \\ 0 \end{pmatrix}^T \begin{pmatrix} x_1 \\ x_2^+ \\ x_2^- \\ z \end{pmatrix} + \frac{1}{2} \begin{pmatrix} x_1 \\ x_2^+ \\ x_2^- \\ z \end{pmatrix}^T \begin{pmatrix} Q_1 & 0 & 0 & 0 \\ 0 & Q_2 & -Q_2 & 0 \\ 0 & -Q_2 & Q_2 & 0 \\ 0 & 0 & 0 & 0 \end{pmatrix} \begin{pmatrix} x_1 \\ x_2^+ \\ x_2^- \\ z \end{pmatrix}$$

subject to

$$\begin{bmatrix} A_{11} & A_{12} & -A_{12} & 0 \\ A_{21} & A_{22} & -A_{22} & -I \end{bmatrix} \begin{bmatrix} x_1 \\ x_2^+ \\ x_2^- \\ z \end{bmatrix} = \begin{bmatrix} b_1 \\ b_2 \end{bmatrix}, \begin{bmatrix} x_1 \\ x_2^+ \\ x_2^- \\ z \end{bmatrix} \geq 0$$

and this is a quadratic program in normal form.

We shall now set up the dual program for the quadratic program in normal form (P). It is (c.f. 4.1):

$$(D_o) \qquad \text{Maximize } \varphi_o(y) := \inf_{x \geq 0} \left(c^T x + \frac{1}{2} x^T Q x + y^T (b - Ax)\right)$$

$$\text{on } N_o := \{y \in \mathbb{R}^m : \varphi_o(y) > -\infty\}.$$

One cannot be entirely satisfied with this formulation of the dual program, since $\varphi_o(y)$ can only be found by solving a quadratic program. Before we reformulate (D_o) we prove an existence theorem and a necessary and sufficient optimality condition for quadratic programs.

4.4.1 Theorem (Existence Theorem): Suppose given a quadratic program in normal form

$$(P) \qquad \text{Minimize } f(x) := c^T x + \frac{1}{2} x^T Q x$$

$$\text{on } M := \{x \in \mathbb{R}^n : Ax = b, \ x \geq 0\}.$$

Suppose $Q \in \mathbb{R}^{n \times n}$ is positive semidefinite. If (P) is feasible, i.e. $M \neq \emptyset$, and

$$\inf (P) = \inf_{x \in M} f(x) > -\infty,$$

then (P) has a solution $\bar{x}$.

Proof: In what follows $|\ |$ is again the euclidean norm in $\mathbb{R}^n$. Since $M \neq \emptyset$ there exists a $\rho_o > 0$ with

$$M_\rho := \{x \in M : |x| \leq \rho\} \neq \emptyset \text{ for all } \rho \geq \rho_o.$$

For $\rho \geq \rho_o$ M_ρ is nonempty and compact and thus the problem

(P_ρ) Minimize $f(x)$ on M_ρ

is solvable. The value min (P_ρ) of (P_ρ) is monotone nonincreasing as a function of ρ, min $(P_\rho) \geq \inf (P)$ and

$$\lim_{\rho\to\infty} \min (P_\rho) = \inf (P).$$

Among all solutions of the problem (P_ρ) (which is not in general uniquely solvable) let $x_\rho \in M_\rho$ be an element of minimal euclidean norm (x_ρ is even uniquely determined by this requirement, but we shall make no use of this fact). The rest of the proof falls into two parts:

1) There exists a $\rho^* \geq \rho_o$ with $|x_{\rho^*}| < \rho^*$.

2) min $(P_{\rho^*}) = \inf (P)$, i.e. x_{ρ^*} is a solution of (P).

Proof of 1): Suppose 1) were false, i.e. $|x_\rho| = \rho$ for all $\rho \geq \rho_o$. In particular we would have $|x_k| = k$ for all $k \in \mathbb{N}$ with $k \geq \rho_o$. Let $y_k := x_k / k$. We can choose a convergent subsequence of $\{y_k\}$. To avoid unnecessary writing we assume $\{y_k\}$ itself converges; let $y := \lim_{k\to\infty} y_k$. Then $y \geq 0$, $|y| = 1$ and $Ay = 0$ since $Ay_k = b/k$. Furthermore $Qy = 0$ and $c^T y = 0$. For: if $k \geq \rho_o$ then

$$-\infty < \inf (P) \leq f(x_k) = kc^T y_k + \frac{1}{2} k^2 y_k^T Q y_k \leq f(x_{\rho_o})$$

and thus

$$c^T y_k/k + \frac{1}{2} y_k^T Q y_k \leq f(x_{\rho_o})/k^2.$$

Letting k tend to ∞ we get $y^T Q y \leq 0$. Since Q is positive semidefinite it follows $Qy = 0$ (proof?). $c^T y \leq 0$ follows from

$$c^T y_k \leq c^T y_k + \frac{1}{2} k y_k^T Q y_k \leq f(x_{\rho_o})/k$$

by letting k tend to ∞.

From $x_k + ty \in M$ for all $t \geq 0$ and

$$-\infty < \inf (P) \leq f(x_k+ty) = f(x_k) + tc^Ty$$

it follows $c^Ty \geq 0$, and hence $c^Ty = 0$.

We shall now show that there is a $t_o > 0$ such that

i) $x_k - ty \in M$

ii) $|x_k-ty| < |x_k|$

for all $t \in (0,t_o)$ and all sufficiently large k. Together with the fact that $f(x_k-ty) = f(x_k)$ (since $c^Ty = 0$, $Qy = 0$) this gives a contradiction to the assumption that x_k was to be a solution of (P_k) with minimal norm.

i) Let $J_o := \{j \in \{1,\dots,n\} : y_j = 0\}$, where $y = (y_1,\dots,y_n)^T$. (Please do not confuse the jth component y_j of the vector y and the jth member of the sequence $\{y_k\}$!). Since $y \geq 0$, $|y| = 1$ there exists an $\varepsilon > 0$ with $0 < \varepsilon \leq y_j \leq 1$ for all $j \notin J_o$. Since $y_k \to y$ we have $(y_k)_j \geq \varepsilon/2$ for all $j \notin J_o$ and all sufficiently large k and thus $(x_k-ty)_j \geq \rho_o\varepsilon/2 - t \geq 0$ for all $t \in (0,\rho_o\varepsilon/2]$ and all sufficiently large $k \geq \rho_o$ and for these t and k we have $x_k - ty \in M$.

ii) Since $|y| = 1$ we have $y_k^Ty \geq \frac{1}{2}$ for all sufficiently large k. Thus

$$\begin{aligned} |x_k-ty|^2 &= |x_k|^2 - 2tky_k^Ty + t^2|y|^2 \\ &\leq |x_k|^2 - 2t\rho_o/2 + t^2 \\ &< |x_k|^2 \quad \text{for all } t \in (0,\rho_o) \end{aligned}$$

and all sufficiently large $k \geq \rho_o$.

Proof of 2): Suppose $\inf (P) < \min (P_{\rho*})$. Then there exists an $x_o \in M$ with $f(x_o) < f(x_{\rho*})$. For sufficiently small $t > 0$ we have

$$x^* := (1-t)x_{\rho*} + tx_o \in M_{\rho*}$$

since $|x_{\rho*}| < \rho^*$ (c.f. 1)). But then (f is convex!)

$$f(x^*) \leq (1-t)f(x_{\rho*}) + tf(x_o) < f(x_{\rho*}),$$

a contradiction to the assumption that $x_{\rho*}$ is a solution of $(P_{\rho*})$.

Remark: The statement of the Existence Theorem 4.4.1 is still true if one only assumes symmetry of Q, as one can learn by reading BLUM-OETTLI [8, p. 122ff]. We have essentially copied their beautiful proof.

In the following theorem we give a necessary and sufficient optimality condition for a quadratic program in normal form. It anticipates a more general assertion (c.f. Theorem 5.3.4).

4.4.2 Theorem: Suppose given a quadratic program in normal form

(P) Minimize $f(x) := c^Tx + \frac{1}{2}x^TQx$ on

$$M := \{x \in \mathbb{R}^n : Ax = b,\ x \geq 0\}.$$

Let $Q \in \mathbb{R}^{n\times n}$ be symmetric and positive semidefinite. Then $\bar{x} \in M$ is a solution of (P) if and only if there exists a $\bar{y} \in \mathbb{R}^m$ with

i) $c + Q\bar{x} - A^T\bar{y} \geq 0$

ii) $\bar{x}^T(c+Q\bar{x}-A^T\bar{y}) = 0$

Proof: 1) Let $\bar{x} \in M$ be a solution of (P). Let

$$\bar{J}_o := \{j \in \{1,\ldots,n\} : \bar{x}_j = 0\} \text{ and}$$

$$F(M;\bar{x}) := \{h \in \mathbb{R}^n : Ah = 0,\ h_j \geq 0 \text{ for } j \in \bar{J}_o\}.$$

One could call $F(M;\bar{x})$ the set of feasible directions at $\bar{x}$, for it is precisely the set of directions h for which $\bar{x} + th$ is still feasible for all sufficiently small $t > 0$. By the optimality of $\bar{x}$ we have

$$\frac{1}{t}\,(f(\bar{x}+th)-f(\bar{x})) = (c+Q\bar{x})^T h + \frac{t}{2}\, h^T Q h \geq 0$$

for all $h \in F(M;\bar{x})$ and all sufficiently small $t > 0$; letting t tend to 0 from above we get

$$(c+Q\bar{x})^T h \geq 0 \text{ for all } h \in F(M;\bar{x}).$$

From this inequality it follows:

$$\begin{bmatrix} A \\ -A \\ e^{jT} \end{bmatrix} h \geq 0 \ (j \in \bar{J}_o), \quad (c + Q\bar{x})^T h < 0$$

(e^j = jth unit vector in $\mathbb{R}^n$) has no solution $h \in \mathbb{R}^n$. The FARKAS lemma 2.2.1 gives the existence of a $\bar{y} \in \mathbb{R}^m$ and $\bar{\lambda}_j \geq 0$ ($j \in \bar{J}_o$) with

$$A^T\bar{y} + \sum_{j \in J_o} \bar{\lambda}_j e^j = c + Q\bar{x}.$$

But then

i) $$c + Q\bar{x} - A^T\bar{y} = \sum_{j \in \bar{J}_o} \bar{\lambda}_j e^j \geq 0$$

ii) $$\bar{x}^T(c+Q\bar{x}-A^T\bar{y}) = \sum_{j \in \bar{J}_o} \bar{\lambda}_j \bar{x}_j = 0$$

2) For $\bar{x} \in M$ let there be a $\bar{y} \in \mathbb{R}^m$ with i), ii). Let $x \in M$ be arbitrary. Then

$$\begin{aligned} f(x) - f(\bar{x}) &= (c+Q\bar{x})^T(x-\bar{x}) + \frac{1}{2}\,(x-\bar{x})^T Q(x-\bar{x}) \\ &\geq (c+Q\bar{x})^T(x-\bar{x}) \\ &\geq (A^T\bar{y})^T(x-\bar{x}) \\ &= \bar{y}^T A(x-\bar{x}) = 0, \end{aligned}$$

i.e. $\bar{x}$ is a solution of (P).

Now we come to the promised reformulation of the dual program

(D_o) Maximize $\varphi_o(y) := \inf_{x\geq 0} (c^Tx + \frac{1}{2} x^TQx + y^T(b-Ax))$

$$\text{on } N_o := \{y \in \mathbb{R}^m : \varphi_o(y) > -\infty\}.$$

$\varphi_o(y)$ is the infimum of the LAGRANGE function

$$L(x,y) = c^Tx + \frac{1}{2} x^TQx + y^T(b-Ax)$$

on the nonnegative orthant $x \geq 0$. From the Existence Theorem 4.4.1 we know $\varphi_o(y) > -\infty$ if and only if the problem of minimizing $L(x,y)$ on $x \geq 0$ has a solution $\hat{x}$, which by Theorem 4.4.2 is equivalent to

$$c + Q\hat{x} - A^Ty \geq 0 \text{ and } \hat{x}^T(c+Q\hat{x}-A^Ty) = 0$$

and thus $\varphi_o(y) = L(\hat{x},y) = b^Ty - \frac{1}{2} \hat{x}^TQ\hat{x}$. Thus we may finally define as dual program to

(P) Minimize $f(x) := c^Tx + \frac{1}{2} x^TQx$

$$\text{on } M := \{x \in \mathbb{R}^n : Ax = b,\ x \geq 0\}$$

the program

(D) Maximize $\varphi(x,y) := b^Ty - \frac{1}{2} x^TQx$

$$\text{on } N := \{(x,y) \in \mathbb{R}^n \times \mathbb{R}^m : c+Qx-A^Ty \geq 0, x \geq 0\}$$

and demonstrate the equivalence of (D) with

(D_o) Maximize $\varphi_o(y) := \inf_{x\geq 0} (c^Tx + \frac{1}{2} x^TQx + y^T(b-Ax))$

$$\text{on } N_o := \{y \in \mathbb{R}^m : \varphi_o(y) > -\infty\}.$$

For: i) $y \in N_o \iff (x,y) \in N$ with an $x \in \mathbb{R}^n$.

ii) $\varphi_o(y) \geq \varphi(x,y)$ for all $(x,y) \in N$. For every $y \in N_o$ there is an $\hat{x} \in \mathbb{R}^n$ with $(\hat{x},y) \in N$ and $\varphi_o(y) = \varphi(\hat{x},y)$. Thus: $N_o \neq \emptyset$

if and only if $N \neq \emptyset$, and $\bar{y} \in N_o$ is a solution of (D_o) if and only if $(\bar{x},\bar{y}) \in N$ is a solution of (D) for some $\bar{x} \in \mathbb{R}^n$.

Remarks: 1) If one transforms the general quadratic program

$$\text{Minimize } c_1^T x_1 + \frac{1}{2} x_1^T Q_1 x_1 + c_2^T x_2 + \frac{1}{2} x_2^T Q_2 x_2 \text{ subject to}$$

$$\left.\begin{array}{l} A_{11}x_1 + A_{12}x_2 = b_1 \\ A_{21}x_1 + A_{22}x_2 \geq b_2 \end{array}\right. , \ x_1 \geq 0$$

into normal form and takes the associated dual program, then one gets

$$\text{Maximize } b_1^T y_1 + b_2^T y_2 - \frac{1}{2} x_1^T Q_1 x_1 - \frac{1}{2} x_2^T Q_2 x_2 \text{ subject to}$$

$$\left.\begin{array}{l} A_{11}^T y_1 + A_{21}^T y_2 \leq c_1 + Q_1 x_1 \\ A_{12}^T y_1 + A_{22}^T y_2 = c_2 + Q_2 x_2 . \end{array}\right. , \ x_1 \geq 0, \ y_2 \geq 0$$

2) Suppose given the primal program (P) in normal form. The associated dual program

$$\text{Minimize } \frac{1}{2} x^T Q x - b^T y \text{ subject to}$$

$$Qx - A^T y \geq - c, \ x \geq 0$$

has a dual program by 1):

$$\text{Minimize } c^T x + \frac{1}{2} z^T Q z \text{ subject to}$$

$$Ax = b, \ x \geq 0$$

$$Qx \leq Qz, \ z \geq 0.$$

But this problem is equivalent to the primal program (P) as one readily sees. Double dualization thus leads us back to the original program, just as in the linear case.

The statement of the following strong duality theorem could be deduced from Theorem 4.3.1 and the Existence Theorem 4.4.1 by showing that

$$\Lambda = \{(b-Ax, c^T x + \frac{1}{2} x^T Qx+r) : x \geq 0, r \geq 0\}$$

is closed. We prefer to give a direct proof.

<u>4.4.3</u> <u>Theorem</u> (Duality Theorem for Quadratic Programs): Suppose given a quadratic program in normal form

(P) $\quad$ Minimize $f(x) := c^T x + \frac{1}{2} x^T Qx$

$$\text{on } M := \{x \in \mathbb{R}^n : Ax = b, \ x \geq 0\}$$

with symmetric, positive semidefinite $Q \in \mathbb{R}^{n\times n}$ and the associated dual program

(D) $\quad$ Maximize $\varphi(x,y) := b^T y - \frac{1}{2} x^T Qx$

$$\text{on } N := \{(x,y) \in \mathbb{R}^n \times \mathbb{R}^m : c + Qx - A^T y \geq 0, \ x \geq 0\}.$$

Then one has:

i) $(x,y) \in N$, $z \in M \Rightarrow \varphi(x,y) \leq f(z)$ (weak duality).

ii) (P) feasible, (D) feasible $\Rightarrow$ (P) has a solution $\bar{x} \in M$ and (D) a solution $(\bar{x},\bar{y}) \in N$ and $\varphi(\bar{x},\bar{y}) = f(\bar{x})$, i.e. max (D) = min (P).

iii) (D) feasible, (P) not feasible $\Rightarrow$ sup (D) $= +\infty$.

iv) (P) feasible, (D) not feasible $\Rightarrow$ inf (P) $= -\infty$.

<u>Proof</u>: i)

$$\begin{aligned} \varphi(x,y) &= b^T y - \frac{1}{2} x^T Qx \\ &= (A^T y)^T z - \frac{1}{2} x^T Qx \\ &\leq (c+Qx)^T z - \frac{1}{2} x^T Qx \\ &= c^T z + \frac{1}{2} z^T Qz - \frac{1}{2} (x-z)^T Q(x-z) \\ &\leq f(z). \end{aligned}$$

ii) If (P) and (D) are feasible then from i) it follows

$$-\infty < \sup(D) \leq \inf(P).$$

From the Existence Theorem 4.4.1 it follows (P) has a solution $\bar{x} \in M$. By Theorem 4.4.2 there exists a $\bar{y} \in \mathbb{R}^m$ such that $(\bar{x},\bar{y}) \in N$ and

$$\bar{x}^T(c+Q\bar{x}-A^T\bar{y}) = 0$$

But then $\varphi(\bar{x},\bar{y}) = f(\bar{x})$ and the assertion follows.

iii) By assumption $N \neq \emptyset$, $M = \emptyset$. Since $Ax = b$ has no nonnegative solution, it follows from the FARKAS lemma 2.2.1 that there is a $u \in \mathbb{R}^m$ with $A^Tu \geq 0$, $b^Tu < 0$. Let $(x,y) \in N$. Then $(x,y-tu) \in N$ for all $t \geq 0$ and $\varphi(x,y-tu) = \varphi(x,y) - tb^Tu$. Letting t tend to $+\infty$ we get $\sup(D) = +\infty$.

iv) can be proved by appealing to symmetry or directly using the FARKAS lemma.

The Existence Theorem and the Strong Duality Theorem naturally hold for a general quadratic program and its dual as well. The statement corresponding to Theorem 4.4.2 is formulated explicitly in the following corollary (proof?).

<u>4.4.4 Corollary:</u> Suppose given the quadratic program

(P) Minimize $f(x_1,x_2) := c_1^Tx_1 + \frac{1}{2}x_1^TQ_1x_1 + c_2^Tx_2 + \frac{1}{2}x_2^TQ_2x_2$

$$\text{on } M := \left\{(x_1,x_2) : \begin{array}{l} A_{11}x_1 + A_{12}x_2 = b_1 \\ A_{21}x_1 + A_{22}x_2 \geq b_2 \end{array},\ x_1 \geq 0\right\}.$$

Here Q_1, Q_2 shall be symmetric and positive semidefinite. Then $(\bar{x}_1,\bar{x}_2) \in M$ is a solution of (P) if and only if there exists a $(\bar{y}_1,\bar{y}_2)$ with

$$\text{i)} \quad \begin{aligned} A_{11}^T\bar{y}_1 + A_{21}^T\bar{y}_2 &\le c_1 + Q_1\bar{x}_1 \\ A_{12}^T\bar{y}_1 + A_{22}^T\bar{y}_2 &= c_2 + Q_2\bar{x}_2 \end{aligned} \quad , \ \bar{y}_2 \ge 0$$

$$\text{ii)} \quad \bar{x}_1^T(c_1+Q_1\bar{x}_1-A_{11}^T\bar{y}_1-A_{21}^T\bar{y}_2) = 0$$

$$\bar{y}_2^T(A_{21}\bar{x}_1+A_{22}\bar{x}_2-b_2) = 0.$$

With the aid of Cor. 4.4.4 one can discuss the uniqueness of solutions of a general quadratic program. For if $(\bar{x}_1,\bar{x}_2)$ is a solution and (x_1,x_2) is feasible, then

$$\begin{aligned} & f(x_1,x_2) - f(\bar{x}_1,\bar{x}_2) \\ & = (c_1+Q_1\bar{x}_1)^T(x_1-\bar{x}_1) + \tfrac{1}{2}(x_1-\bar{x}_1)^TQ_1(x_1-\bar{x}_1) \\ & \quad + (c_2+Q_2\bar{x}_2)^T(x_2-\bar{x}_2) + \tfrac{1}{2}(x_2-\bar{x}_2)^TQ_2(x_2-\bar{x}_2) \\ & \ge (A_{11}^T\bar{y}_1+A_{21}^T\bar{y}_2)^T(x_1-\bar{x}_1) + \tfrac{1}{2}(x_1-\bar{x}_1)^TQ_1(x_1-\bar{x}_1) \\ & \quad + (A_{12}^T\bar{y}_1+A_{22}^T\bar{y}_2)^T(x_2-\bar{x}_2) + \tfrac{1}{2}(x_2-\bar{x}_2)Q_2(x_2-\bar{x}_2) \\ & \ge \tfrac{1}{2}(x_1-\bar{x}_1)^TQ_1(x_1-\bar{x}_1) + \tfrac{1}{2}(x_2-\bar{x}_2)^TQ_2(x_2-\bar{x}_2). \end{aligned}$$

Thus a solution is unique if Q_1 and Q_2 are positive definite (if all variables are nonnegative, i.e. $A_{12} = 0$, $A_{22} = 0$, then of course the positive definiteness of Q_1 suffices).

<u>Example</u>: We wish to return to the example 2) at the beginning of the section. The determination of an optimal dose of medication in a simple compartment model led us to the quadratic program

$$(*) \qquad \text{Mimimize } \tfrac{1}{2}\,\delta^T\delta \text{ subject to } \delta \ge 0,\ \Phi\delta \ge me.$$

If (*) is feasible, then there exists a unique solution $\bar{\delta}$ (the existence follows e.g. from the Existence Theorem 4.4.1, the uniqueness from the remark above) and by Cor. 4.4.4 it is characterized by the existence of a $\bar{y} \in \mathbb{R}^{N+1}$ with

$$\bar{\delta} - \Phi^T\bar{y} \geq 0,\ \bar{y} \geq 0 \quad \text{and}$$

$$\bar{\delta}^T(\bar{\delta}-\Phi^T\bar{y}) = 0,\ \bar{y}^T(\Phi\bar{\delta}-me) = 0.$$

If one considers except for an initial or adjustment phase only equidistant τ_i, i.e.

$$\tau_o = 0 \quad \text{and} \quad \tau_i := \tau + (i-1)\Delta \quad (i=1,\ldots,N+1)$$

with $\tau,\Delta > 0$, then for practical reasons it seems appropriate to consider doses of the form $\delta = (\delta_o,\delta_1,\ldots,\delta_1)^T$. Then one has the question under what hypotheses a solution $(\bar{\delta}_o,\bar{\delta}_1)$ of

(**) $$\text{Minimize } \frac{1}{2}\,\delta_o^2 + \frac{N}{2}\,\delta_1^2 \text{ subject to } \delta_o,\delta_1 \geq 0,$$

$$\varphi(\tau+i\Delta)\,\delta_o + \sum_{j=1}^{i} \varphi(j\Delta)\,\delta_1 \geq m \quad (i=0,\ldots,N)$$

gives a solution of (*) of the form $\bar{\delta} = (\bar{\delta}_o,\bar{\delta}_1,\ldots,\bar{\delta}_1)^T$. Here

$$\varphi(t) = \frac{k_1}{k_2-k_1}\,(e^{-k_1 t} - e^{-k_2 t})$$

with positive distinct constants k_1,k_2; $\Phi \in \mathbb{R}^{(N+1)\times(N+1)}$ is given for equidistant τ_i by

$$\Phi = \begin{pmatrix} \varphi(\tau) & 0 & 0 & \cdots & 0 \\ \varphi(\tau+\Delta) & \varphi(\Delta) & 0 & \cdots & 0 \\ \varphi(\tau+2\Delta) & \varphi(2\Delta) & \varphi(\Delta) & \cdots & 0 \\ \vdots & \vdots & \vdots & & \vdots \\ \varphi(\tau+N\Delta) & \varphi(N\Delta) & \varphi((N-1)\Delta) & \cdots & \varphi(\Delta) \end{pmatrix}$$

We shall not discuss this example further, but rather refer the reader to PIERCE-SCHUMITZKY [64].

4.5 Literature

To 4.2, 4.3: The presentation in these two sections was

most influenced by LUENBERGER [53], KRABS [45], BAZARAA-SHETTY [3], ROCKAFELLAR [68], WETS [77], VAN SLYKE-WETS [75]. There is a great deal of literature on the Jung inequality, c.f. e.g. JUNG [39], BLUMENTHAL-WAHLIN [9] (with several historical remarks), VEBLUNSKY [76]. We adopted the proof of the Jung inequality from IOFFE-TICHOMIROV [37]. A proof of the alternation theorem 4.3.11 can be found in practically any book on approximation theory. As one example we mention CHENEY [13].

To 4.4: The duality theory for quadratic programs which began in 1960 with papers by DORN and COTTLE can be found in such textbooks as MANGASARIAN [57], COLLATZ-WETTERLING [14], BLUM-OETTLI [8], STOER-WITZGALL [71].

§ 5 NECESSARY OPTIMALITY CONDITIONS

5.1 GATEAUX and FRECHET Differential

In the following sections we shall consider optimization problems of the form

(P) Minimize $f(x)$ on $M := \{x \in X : x \in C, g(x) \in -K\}$.

We shall assume as in the last chapter that $C \subset X$ is convex and $K \subset Y$ a convex cone, but we shall no longer assume that $f : X \to \mathbb{R}$ is convex and $g : X \to Y$ convex with respect to K. Instead we assume that f and g can be "linearized" in a point $\bar{x} \in M$ (for necessary optimality conditions $\bar{x}$ will always be a solution of (P)), so that locally (P) can be associated with a program of a simple sort (convex or even linear). If one thinks of the TAYLOR expansion then the connection between "linearization" and "differentiation" is obvious and this is the reason why the concept of differentiability must be carried over to maps between normed linear spaces. In recent years many different definitions of differentiability have been developed especially for application to optimization problems. We are concerned here with presenting the basic ideas needed for deducing necessary optimality conditions. Therefore we do not go into various possible generalizations, but only present the two most important definitions of differentiability, those of GATEAUX and FRECHET differentiability.

First we introduce an important notation. Let X and Y be normed linear spaces (the norms will always be denoted by $\|\ \|$; from the context it should be clear whether a norm on X or on Y is meant). Let L(X,Y) denote the set of continuous linear maps from X to Y. L(X,Y) is made into a linear space in the canonical fashion. If one moreover defines

$$\|T\| := \sup_{x \neq 0} \frac{\|Tx\|}{\|x\|}$$

for $T \in L(X,Y)$, then $(L(X,Y), \|\ \|)$ is a normed linear space.

5.1.1 Definition: Let X,Y be normed linear spaces and $g : X \to Y$ a map, $x \in X$.

i) If for every $h \in X$ the limit

$$g'(x;h) := \lim_{t \to 0+} \frac{1}{t} \{g(x+th)-g(x)\}$$

exists, then $g'(x;\cdot) : X \to Y$ is the GATEAUX variation (or simply G-variation) of g at x.

ii) If g has a G-variation $g'(x;\cdot) \in L(X,Y)$ at x, then it is called a G-differential and is denoted by $g'(x)$. In this case g is said to be G-differentiable at x.

iii) g is called continuously G-differentiable at x if g is G-differentiable at every point of a neighborhood U of x and the map $z \to g'(z)$ from U to $L(X,Y)$ is continuous at x (i.e. if for every $\varepsilon > 0$ ther exists a $\delta = \delta(\varepsilon) > 0$ with:

$$z \in B[x;\delta] \cap U \Rightarrow \| g'(z)-g'(x) \| \leq \varepsilon).$$

Remarks: 1) In Theorem 3.3.2 we showed: If $f \in \text{Conv}\,(X)$, i.e. $f : X \to \mathbb{R}$ is convex, then the G-variation $f'(x)$ exists. By Theorem 3.3.3 f' is convex but in general it is not linear and continuous and hence not a G-differential.

2) If the G-variation $g'(x;\cdot)$ is linear (e.g. a G-differential) then obviously

$$g'(x;h) = \lim_{t \to 0} \frac{1}{t} \{g(x+th)-g(x)\},$$

i.e. not just the one-sided limit exists.

The connection between the G-differential and the subdifferential of a convex real-valued function is given by:

5.1.2 Lemma: Let X be a normed linear space and let $f \in \text{Conv}\,(X)$ be G-differentiable at $x \in X$. Then

$$\partial f(x) = \partial f(x) \cap X^* = \{f'(x)\}.$$

<u>Proof</u>: 1) We have $f'(x)(z-x) \le f(z) - f(x)$ for all $z \in X$, as we showed in Theorem 3.3.2. Thus $f'(x) \in \partial f(x)$.

2) Suppose $x^* \in \partial f(x)$, i.e. x^* linear with $\langle x^*, z-x\rangle \le f(z)-f(x)$ for all $z \in X$. Then $\langle x^*, h\rangle \le \frac{1}{t}\{f(x+th)-f(x)\}$ for all $t > 0$, and hence $\langle x^*, h\rangle \le f'(x)h$ for all $h \in X$. By replacing h by $-h$ one gets $\langle x^*, h\rangle = f'(x)h$ for all $h \in X$ and thus the claim.

If $g : X \to Y$ is G-differentiable at every point of $[x,x+h]$ $(= \{x + th : t \in [0,1]\})$, then it is in general true that

$$g(x+h) - g(x) \neq g'(z)h \text{ for every } z \in [x,x+h],$$

i.e. the mean value theorem of analysis is not directly translatable (Example?). However, we do have:

<u>5.1.3 Theorem</u>: Let X,Y be normed linear spaces and $g : X \to Y$ G-differentiable at every point of $[x,x+h]$. Then

i) $$\| g(x+h)-g(x) \| \le \sup_{0<t<1} \| g'(x+th)\| \, \| h \|.$$

ii) If $T \in L(X,Y)$ then

$$\| g(x+h)-g(x)-Th \| \le \sup_{0<t<1} \| g'(x+th)-T \| \, \| h \|.$$

In particular

$$\| g(x+h)-g(x)-g'(x)h \| \le \sup_{0<t<1} \| g'(x+th)-g'(x) \| \, \| h \|.$$

<u>Proof</u>: i) By Corollary 3.2.7 there exists an $l \in Y^*$ with $\| l \| = 1$ and $\langle l, g(x+h)-g(x)\rangle = \| g(x+h)-g(x) \|$. Now define

$$\varphi : [0,1] \to \mathbb{R} \text{ by } \varphi(t) = \langle l, g(x+th)\rangle.$$

Then φ is differentiable on $[0,1]$ and $\varphi'(t) = \langle l, g'(x+th)h\rangle$ (proof?). From the mean value theorem of differential calculus we get the existence of a $t_0 \in (0,1)$ with $\varphi(1) - \varphi(0) = \varphi'(t_0)$.

Thus

$$\| g(x+h)-g(x) \| = \varphi(1) - \varphi(0) = \varphi'(t_o) = \langle l,g'(x+t_oh)h\rangle$$
$$\leq \| g'(x+t_oh) \| \| h \|$$
$$\leq \sup_{0<t<1} \| g'(x+th) \| \| h \| .$$

ii) Apply i) to $G(x) := g(x) - Tx$.

The concept of G-differentiablity does not suffice for nonlinear optimization problems, in particular for problems with nonlinear equality constraints. Therefore:

5.1.4 Definition: Let X,Y be normed linear spaces and $g : X \to Y$ a map, $x \in X$. Then g is called FRECHET differentiable at x (or simply F-differentiable) if $g'(x) \in L(X,Y)$ exists with

$$\lim_{\| h \| \to 0} \frac{\| g(x+h)-g(x)-g'(x)h \|}{\| h \|} = 0,$$

in which case $g'(x)$ is called the F-differential of F-derivative of g at x. g is continuously F-differentiable at x if g is F-differentiable at every point of a neighborhood U of x and the map $z \to g'(z)$ from U to $L(X,Y)$ is continuous.

Remarks: 1) If g is F-differentiable at x then the F-differential $g'(x) \in L(X,Y)$ is uniquely determined. For suppose $g_1'(x)$, $g_2'(x) \in L(X,Y)$ were two F-differentials of g at x. For $\varepsilon > 0$ there exist $\delta_1, \delta_2 > 0$ with

$$\| h \| \leq \delta_i \Rightarrow \| g(x+h)-g(x)-g_i'(x)h \| \leq \frac{\varepsilon}{2} \| h \| \quad (i=1,2).$$

First for $\| h \| \leq \delta := \min(\delta_1,\delta_2)$ and then for all h we have

$$\| (g_1'(x)-g_2'(x))h \| \leq \varepsilon \| h \| .$$

From this it follows that $\| g_1'(x)-g_2'(x) \| \leq \varepsilon$ and letting ε tend to 0 we get $g_1'(x) = g_2'(x)$.

2) If g is F-differentiable at x, then g is continuous at x (proof?).

3) If g is F-differentiable at x, then g is also G-differentiable at x and the F-differential and G-differential agree (proof?). Thus Theorem 5.1.3 also holds for F-differentiable functions.

4) It is trivial to show: if $g_1, g_2 : X \to Y$ are F-differentiable at x then so is $\alpha_1 g_1 + \alpha_2 g_2$ $(\alpha_1, \alpha_2 \in \mathbb{R})$ and

$$(\alpha_1 g_1 + \alpha_2 g_2)'(x) = \alpha_1 g_1'(x) + \alpha_2 g_2'(x).$$

Somewhat more difficult is the proof of the chain rule: Let X_1, X_2, X_3 be normed linear spaces, $g_1 : X_1 \to X_2$ F-differentiable at $x \in X_1$ and $g_2 : X_2 \to X_3$ F-differentiable at $g_1(x)$. Then $g_2 \circ g_1 : X_1 \to X_3$ is F-differentiable at x and

$$(g_2 \circ g_1)'(x) = g_2'(g_1(x)) \circ g_1'(x) \qquad \text{(proof?)}.$$

We close this section by giving several examples that are important for applications.

<u>Examples</u>: 1) Let $f : \mathbb{R}^n \to \mathbb{R}$ be continuously partially differentiable at $x \in \mathbb{R}^n$. Then f is continuously F-differentiable and

$$f'(x)h = \sum_{j=1}^{n} \frac{\partial f}{\partial x_j}(x)h_j = \nabla f(x)^T h,$$

where

$$\nabla f(x) = \left(\frac{\partial f}{\partial x_1}(x), \ldots, \frac{\partial f}{\partial x_n}(x)\right)^T$$

is the <u>gradient of f</u> at x. This is proved in analysis, c.f. e.g. APOSTOL [1, p. 118] or LUENBERGER [53, p. 173]. If $g : \mathbb{R}^n \to \mathbb{R}^m$ is continuously partially differentiable at $x \in \mathbb{R}^n$, i.e. the partial derivatives $\frac{\partial g_i}{\partial x_j}$ exist in a neighborhood of x and are continuous at x, then g is continuously F-differentiable at x and

$$g'(x)h = \left(\sum_{j=1}^{n} \frac{\partial g_i}{\partial x_j} (x) h_j \right).$$

Hence $g'(x)$ can be identified with the Jacobi matrix

$$\left(\frac{\partial g_i}{\partial x_j} (x) \right)_{1 \leq i, j \leq n}$$

2) In the calculus of variations objective functions of the form

$$I(x) := \int_{t_o}^{t_1} f(x(t), \dot{x}(t), t) dt$$

occur. Here one must ask oneself on which normed linear space I should be defined. Here we shall only consider the simplest (not always adequate) case and merely give some hints about a more general one. Let $X := C_n^1[t_o, t_1]$ $(= C^1([t_o, t_1], \mathbb{R}^n))$ be the linear space of continuously differentiable functions from $[t_o, t_1]$ to $\mathbb{R}^n$. X is made into a normed linear space by defining

$$\| x \| := \max (\| x \|_\infty, \| \dot{x} \|_\infty)$$

for $x \in X$, where

$$\| x \|_\infty := \max_{t \in [t_o, t_1]} |x(t)|.$$

We want to show:

If $f : \mathbb{R}^n \times \mathbb{R}^n \times [t_o, t_1] \to \mathbb{R}$ is continuous and continuously partially differentiable in x and $\dot{x}$ (resp. in the components of $x, \dot{x}$), i.e. if the gradient

$$\nabla_x f(x, \dot{x}, t) = \left(\frac{\partial f}{\partial x_1} (x, \dot{x}, t), \ldots, \frac{\partial f}{\partial x_n} (x, \dot{x}, t) \right)^T$$

of f with respect to x and correspondingly $\nabla_{\dot{x}} f(x, \dot{x}, t)$ exist and if these gradients are continuous on $\mathbb{R}^n \times \mathbb{R}^n \times [t_o, t_1]$, then $I : X \to \mathbb{R}$ is continuously F-differentiable at every $x \in X$ and

$$I'(x)h = \int_{t_o}^{t_1} \{\nabla_x f(x(t),\dot{x}(t),t)^T h(t) + \nabla_{\dot{x}} f(x(t),\dot{x}(t),t)^T \dot{h}(t)\}dt.$$

For: i) The given $I'(x)$ is an element of $L(X,\mathbb{R})$ (resp. X^*), since $I'(x)$ is linear and

$$|I'(x)h| \le \int_{t_o}^{t_1} \{|\nabla_x f(t)||h(t)| + |\nabla_{\dot{x}} f(t)||\dot{h}(t)|\}dt$$

$$\le c\|h\| \text{ for a certain constant } c > 0$$

and that is precisely the continuity of $I'(x)$. Here we have employed the abbreviation $\nabla_x f(t) = \nabla_x f(x(t),\dot{x}(t),t)$; $\nabla_{\dot{x}} f(t)$ is defined analogously.

ii) $$\lim_{\|h\|\to 0} \frac{1}{\|h\|} |I(x+h)-I(x)-I'(x)h| = 0.$$

For: Define

$$Q := \{(p,q,t) \in \mathbb{R}^n \times \mathbb{R}^n \times [t_o,t_1] : |p|,|q| \le \|x\| + 1\}.$$

Since $\nabla_x f$, $\nabla_{\dot{x}} f$ are uniformly continuous on Q there is for a given $\varepsilon > 0$ a $\delta \in (0,1]$ with:

$$(p_1,q_1,t),(p_2,q_2,t) \in Q,\ \max(|p_1-p_2|,|q_1-q_2|) \le \delta \Rightarrow$$

$$\max(|\nabla_x f(p_1,q_1,t)-\nabla_x f(p_2,q_2,t)|,$$

$$|\nabla_{\dot{x}} f(p_1,q_1,t)-\nabla_{\dot{x}} f(p_2,q_2,t)|) \le \frac{\varepsilon}{2(t_1-t_o)} .$$

Let $\|h\| \le \delta$. If $t \in [t_o,t_1]$ and $\theta(t) \in [0,1]$, then

$$(x(t)+\theta(t)h(t),\dot{x}(t)+\theta(t)\dot{h}(t),t) \in Q.$$

Applying the mean value theorem and the definition of δ gives

$$|I(x+h)-I(x)-I'(x)h| \le \varepsilon\|h\|$$

and thus ii) is proved.

iii) I is even continuously F-differentiable at x.
For: Since we have made global assumptions I is F-differentiable at every x. Let $\varepsilon > 0$ be given arbitrarily, Q and δ as in ii). If $\| z-x \| \leq \delta$, then $(z(t),\dot{z}(t),t) \in Q$ and

$$|z(t)-x(t)|, |\dot{z}(t)-\dot{x}(t)| \leq \delta,$$

and thus for arbitrary $h \in X$

$$|(I'(z)-I'(x))h| \leq \varepsilon \| h \|, \text{ i.e. } \| I'(z)-I'(x) \| \leq \varepsilon,$$

which is precisely the continuous F-differentiability of I at x.

If one chooses $X = C^1_{pc,n}[t_o,t_1]$ as range of definition for I, i.e. the linear space of piecewise continuously differentiable functions from $[t_o,t_1]$ to $\mathbb{R}^n$ (with the same norm as in $C^1_n[t_o,t_1]$), then one can argue as above. The same holds for

$$X = W_n^{1,\infty}[t_o,t_1] := \{x \in C_n[t_o,t_1] : \exists\, y \in L_n^\infty[t_o,t_1] \text{ with } x(t) = x(t_o) + \int_{t_o}^{t} y(d)ds\}.$$

Here $L_n^\infty[t_o,t_1]$ is the linear space of measurable functions from $[t_o,t_1]$ to $\mathbb{R}^n$ which are essentially bounded. For $x \in L_n^\infty[t_o,t_1]$ there exists a constant $c > 0$ with $|x(t)| \leq c$ for almost all $t \in [t_o,t_1]$. $L_n^\infty[t_o,t_1]$ is made into a normed linear space by setting

$$\| x \|_\infty := \inf \{c : |x(t)| \leq c \text{ a.e. on } [t_o,t_1]\}.$$

If $x(t) = x(t_o) + \int_{t_o}^{t} y(s)ds$ with $y \in L_n^\infty[t_o,t_1]$, then x is differentiable a.e. on $[t_o,t_1]$ and $\dot{x} = y$ (a.e. equal functions are identified) (c.f. e.g. HEWITT-STROMBERG [34 ,p. 286]). As norm on $W_n^{1,\infty}[t_o,t_1]$ one defines $\| x \| = \max(\| x \|_\infty, \| \dot{x} \|_\infty)$. Then with the same hypotheses as above one can without great difficulty show the F-differentiability of

$$I(x) = \int_{t_o}^{t_1} f(x(t),\dot{x}(t),t)dt \text{ on } W_n^{1,\infty}[t_o,t_1].$$

For many of the functionals and maps occuring in applications one can prove the F-differentiablity along the same lines as above. If only the continuous F-differentiablity at a point is required, then of course local smoothness hypotheses suffice, here for the function f.

5.2 The Theorem of LYUSTERNIK

Of fundamental importance in the theory of necessary optimality conditions is the concept of a tangent cone. We define

5.2.1 Definition: If X is a normed linear space, $M \subset X$ and $\bar{x} \in M$, then

$$T(M;\bar{x}) := \{h \in X : \exists \{t_j\} \subset \mathbb{R}_+ \text{ with } t_j \to 0, \{r_j\} \subset X$$

$$\text{with i)} \quad \bar{x} + t_j h + r_j \in M,$$

$$\text{ii)} \quad \lim_{j\to\infty} \frac{r_j}{t_j} = 0\}$$

is called the tangent cone to M at $\bar{x}$.

Before we investigate properties and examples of tangent cones we shall try to explain why the tangent cone is so important for necessary optimality conditions.

Suppose given the optimization problem

(P) Minimize $f(x)$ on $M \subset X$,

where X is a normed linear space and $f : X \to \mathbb{R}$. Let $\bar{x} \in M$ be a local solution of (P), i.e. suppose there exists a neighborhood U of $\bar{x}$, e.g. a sphere around $\bar{x}$, with $f(\bar{x}) \leq f(x)$ for all $x \in U \cap M$. Let f be F-differentiable at $\bar{x}$. For $h \in T(M;\bar{x})$ we have

$$f'(\bar{x})h = \lim_{j\to\infty} \frac{1}{t_j} \{f(\bar{x}+t_j h+r_j)-f(\bar{x})\} \geq 0.$$

If one can find a (nontrivial) convex cone $L(M;\overline{x}) \subset T(M;\overline{x})$ (with suitable hypotheses this will be possible by the Theorem of LYUSTERNIK), then $\overline{h} = 0$ is a solution of the convex program

$$\text{Minimize } f'(\overline{x})h \text{ on } L(M;\overline{x}).$$

The application of a strong duality theorem of convex optimization leads to the desired necessary optimality conditions.

Remarks: 1) The tangent cone $T(M;\overline{x})$ to M at $\overline{x}$ is a closed cone in X. We leave the proof that $T(M;\overline{x})$ is closed as an exercise (c.f. e.g. KRABS [45, p. 146], BARARAA-SHETTY [3, p. 76]); it is trivial that $T(M;\overline{x})$ is a cone. In general $T(M;\overline{x})$ is not convex:

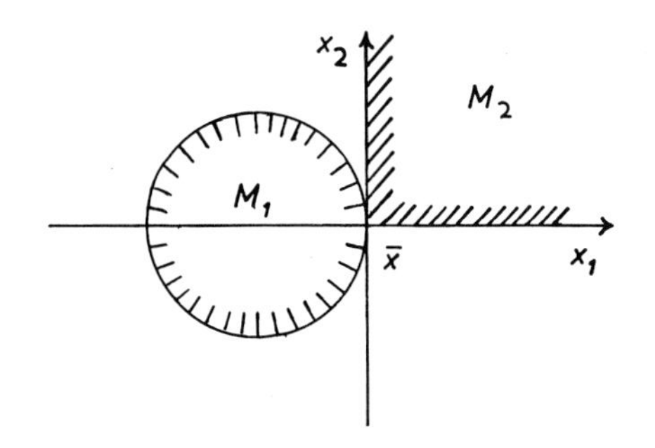

Let $M = M_1 \cup M_2$ with
$M_1 = \{(x_1,x_2):(x_1+1)^2+x_2^2 \leq 1\}$
$M_2 = \{(x_1,x_2) : x_1 \geq 0, x_2 \geq 0\}$
and $\overline{x} = (0,0)$. Then
$T(M;\overline{x}) = T(M_1;\overline{x}) \cup T(M_2;\overline{x})$ with
$T(M_1;\overline{x}) = \{(x_1,x_2) : x_1 \leq 0\}$
$T(M_2;\overline{x}) = \{(x_1,x_2) : x_1,x_2 \geq 0\}$.

$T(M;\overline{x})$ is not convex.

2) Let $M \subset X$ be convex, $\overline{x} \in M$. Further let

$$M(\overline{x}) := \{\lambda(x-\overline{x}) : \lambda \geq 0, x \in M\}$$

($M(\overline{x})$ is precisely the cone generated by $M - \overline{x}$. It is convex, since M resp. $M - \overline{x}$ is convex.) Then $T(M;\overline{x}) = \text{cl } M(\overline{x})$ (proof ?).

Before we can formulate and prove the Theorem of LYUSTERNIK we must prepare some necessary tools. First we remind the reader of some important concepts of elementary functional analysis: A sequence $\{x_k\}$ in a normed linear space X is a Cauchy sequence if for every $\varepsilon > 0$ there is $K(\varepsilon) \in \mathbb{N}$ with $\| x_k - x_l \| \leq \varepsilon$ for all $k,l \geq K(\varepsilon)$. Obviously every convergent sequence is a Cauchy sequence. A normed linear space is complete resp. a <u>Banach space</u> if:

$$\{x_k\} \subset X \text{ Cauchy sequence} \Rightarrow \exists\, x \in X \text{ with } \lim_{k\to\infty} x_k = x.$$

<u>Examples:</u> 1) Every finite dimensional normed linear space is a Banach space.

2) $C_n[t_o,t_1]$, $L_n^\infty[t_o,t_1]$ given the norms

$$\| x \|_\infty := \max_{t\in[t_o,t_1]} |x(t)|$$

resp. $\| x \|_\infty = \inf \{c : |x(t)| \leq c \text{ a.e. on } [t_o,t_1]\}$ are Banach spaces.

3) $$W_n^{1,\infty}[t_o,t_1] = \{x \in C_n[t_o,t_1] : x(t) = x(t_o) + \int_{t_o}^{t} \dot{x}(s)ds \text{ with } \dot{x} \in L_n^\infty[t_o,t_1]\}$$

given the norm $\| x \| = \max (\| x \|_\infty, \| \dot{x} \|_\infty)$ is a Banach space. For: let $\{x_k\} \subset W_n^{1,\infty}[t_o,t_1]$ be a Cauchy sequence. Then $\{x_k\}$, $\{\dot{x}_k\}$ are Cauchy sequences in $C_n[t_o,t_1]$ resp. $L_n^\infty[t_o,t_1]$. Since these are Banach spaces there exist $x \in C_n[t_o,t_1], y \in L_n^\infty[t_o,t_1]$ with

$$\| x_k - x \|_\infty \to 0, \quad \| \dot{x}_k - y \|_\infty \to 0.$$

For fixed $t \in [t_o,t_1]$ the left side of

$$x_k(t) - x_k(t_o) = \int_{t_o}^{t} \dot{x}_k(s)ds$$

converges to $x(t) - x(t_o)$, whereas the right side converges to

$$\int_{t_o}^{t} y(s)ds$$

since

$$\left| \int_{t_o}^{t} \dot{x}_k(s)ds - \int_{t_o}^{t} y(s)ds \right| \leq (t_1 - t_o) \| \dot{x}_k - y \|_\infty .$$

Thus

$$x(t) = x(t_o) + \int_{t_o}^{t} y(s)ds$$

and hence $x \in W_n^{1,\infty}[t_o,t_1]$, $\| x_k - x \| \to 0$. Thus we have demonstrated the convergence of $\{x_k\}$ to an element of $W_n^{1,\infty}[t_o,t_1]$; $W_n^{1,\infty}[t_o,t_1]$ is a Banach space.

A classical theorem of functional analysis is the theorem of BAIRE, which we formulate for our purposes as follows:

5.2.2 Theorem (BAIRE): If X is a Banach space and $A_1, A_2, \ldots$ are closed subsets of X with

$$X = \bigcup_{i=1}^{\infty} A_i,$$

then $\text{int}\,(A_j) \neq \emptyset$ for at least one $j \in \mathbb{N}$.

Proof: See almost any textbook on functional analysis or e. g. LUENBERGER [53, p. 148].

We shall need the following two notational abbreviations.

i) If C is a subset of the normed linear space X and $\bar{x} \in C$, then let

$$C(\bar{x}) := \{\lambda(c-\bar{x}) : \lambda \geq 0,\ c \in C\}.$$

$C(\bar{x})$ is the cone generated by $C - \bar{x}$, i.e. the smallest cone containing $C - \bar{x}$. If C is convex, then $C(\bar{x})$ is also convex. If C is closed, then $C(\bar{x})$ is not necessarily closed (example?).

ii) If C is a subset of the normed linear space X and $\bar{x} \in C$, then let

$$(C-\bar{x})_1 := (C-\bar{x}) \cap B[0;1].$$

If C is convex, then $(C-\bar{x})_1$ is also convex and thus we have:

$$\alpha(C-\bar{x})_1 + \beta(C-\bar{x})_1 = (\alpha+\beta)(C-\bar{x})_1 \text{ for all } \alpha,\beta \geq 0$$

(proof?). If C is closed, then so is $(C-\bar{x})_1$.

The following generalization of the open mapping theorem is due to ZOWE-KURCYUSZ [81] and is the most important ingredient of the proof of the theorem of LYUSTERNIK.

<u>5.2.3 Theorem:</u> Let X,Y be Banach spaces and $C \subset X$, $K \subset Y$ be nonempty, convex and closed. Suppose $T \in L(X,Y)$, $\bar{x} \in C$ and $\bar{y} \in K$. Furthermore suppose

$$\text{(CQ)} \qquad TC(\bar{x}) + K(\bar{y}) = Y.$$

Then there exists a $\rho > 0$ with

$$B[0;\rho] \subset T(C-\bar{x})_1 + (K-\bar{y})_1.$$

Before we prove the theorem we wish to explain why we say this is an open mapping theorem. If in Thm. 5.2.3 we have the special case $C = X$ and $K = \{0\}$, then the statement reads: let X,Y be Banach spaces and $T \in L(X,Y)$ surjective; then there exists a $\rho > 0$ with $B[0;\rho] \subset TB[0;1]$. From this, however, it follows immediately that T is an open mapping, i.e. under T the image of an open set is again open, and this is precisely the assertion of the classical open mapping theorem.

<u>Proof of Theorem 5.2.3:</u> 1) $C(\bar{x}) = \bigcup_{i=1}^{\infty} i(C-\bar{x})_1$. For let $\lambda(c-\bar{x}) \in C(\bar{x})$, i.e. $\lambda \geq 0$ and $c \in C$. Now choose $i \in \mathbb{N}$ so large that $\lambda \leq i$ and

$$\frac{\lambda}{i} \| c-\bar{x} \| \leq 1.$$

Then $\lambda(c-\bar{x}) = i(\frac{\lambda}{i} c+(1- \frac{\lambda}{i})\bar{x}-\bar{x}) \in i(C-\bar{x})_1$. If one defines

$$A_\alpha := \alpha(T(C-x)_1 + (K-\bar{y})_1) \text{ for } \alpha \geq 0,$$

then because of (CQ):

$$Y = TC(\bar{x}) + K(\bar{y}) = \bigcup_{i=1}^{\infty} A_i = \bigcup_{i=1}^{\infty} \text{cl}(A_i).$$

From the Theorem of BAIRE it follows that there is a $j \in \mathbb{N}$

with int $(cl(A_j)) \neq \emptyset$. Take an $a \in$ int $(cl(A_j))$. Because

$$Y = \bigcup_{i=1}^{\infty} cl(A_i)$$

there is an $i \in \mathbb{N}$ with $-a \in cl(A_i)$. Then $-\frac{j}{i} a \in cl(A_j)$. Since A_j and thus $cl(A_j)$ is convex, it follows from Lemma 3.2.1 i) that $(-\frac{j}{i} a, a] \subset int(cl(A_j))$. In particular $0 \in$ int $(cl(A_j))$ and thus also $0 \in$ int $(cl(A_1))$. Hence there exists a $\rho > 0$ with $B[0;2\rho] \subset cl(A_1)$.

2) We have $B[0;\rho] \subset A_1 = T(C-\bar{x})_1 + (K-\bar{y})_1$ with ρ as in 1). For by the choice of ρ we have

$$B[0;\rho] = \frac{1}{2} B[0;2\rho] \subset \frac{1}{2} cl(A_1) \subset \frac{1}{2} A_1 + B[0;\tfrac{1}{2}\rho]$$

and thus for $i = 0,1,2,...$

$$(*) \qquad B[0;(\tfrac{1}{2})^i \rho] = (\tfrac{1}{2})^i B[0;\rho] \subset (\tfrac{1}{2})^{i+1} A_1 + B[0;(\tfrac{1}{2})^{i+1}\rho]$$

Let $y \in B[0;\rho]$ be arbitrary. Applying (*) for $i = 0$ gives:

$$y = \frac{1}{2}(Tu_1+v_1) + r_1 \text{ with } u_1 \in (C-\bar{x})_1,\ v_1 \in (K-\bar{y})_1$$

and $r_1 \in B[0;(\frac{1}{2})^1\rho]$. Repeated application shows:

$$y = Tx_k + z_k + r_k \text{ with } x_k = \sum_{i=1}^{k} (\tfrac{1}{2})^i u_i,\ u_i \in (C-\bar{x})_1, \text{ and}$$

$$z_k = \sum_{i=1}^{k} (\tfrac{1}{2})^i v_i,\ v_i \in (K-\bar{y})_1,\ r_k \in B[0;(\tfrac{1}{2})^k\rho].$$

We have $\{x_k\} \subset (C-\bar{x})_1$, since

$$x_k \in \sum_{i=1}^{k} (\tfrac{1}{2})^i (C-\bar{x})_1 = (1-(\tfrac{1}{2})^k)(C-\bar{x})_1 \subset (C-\bar{x})_1.$$

$\{x_k\}$ is obviously a Cauchy sequence. Since X is a Banach space and $(C-\bar{x})_1$ is closed, there exists an $x \in (C-\bar{x})_1$ with $x_k \to x$. Similarly $\{z_k\} \subset (K-\bar{y})_1$ converges to a $z \in (K-\bar{y})_1$. Since $\| r_k \| \leq (\frac{1}{2})^k\rho$ and $y = Tx_k + z_k + r_k$ the continuity of T gives us finally $y = Tx + z \in T(C-\bar{x})_1 + (K-\bar{y})_1$ and thus the theorem is proved.

We record the following simple corollary of Theorem 5.2.3.

<u>5.2.4 Corollary</u>: Suppose the hypotheses of Theorem 5.2.3 are fulfilled. Let

$$\rho_T := \sup\{\rho > 0 : B[0;\rho] \subset T(C-\bar{x})_1 + (K-\bar{y})_1\}.$$

Then for $L > 1/\rho_T$ and $y \in Y$ there exist

$$\begin{Bmatrix} x \\ z \end{Bmatrix} \in L\|y\| \begin{Bmatrix} (C-\bar{x})_1 \\ (K-\bar{y})_1 \end{Bmatrix} \quad \text{with } y = Tx + z.$$

<u>Remark</u>: In particular let $C = X$ and $K = \{0\}$ in Corollary 5.2.4 and suppose $T \in L(X,Y)$ is surjective (then (CQ) is fulfilled). If one forms the quotient space $X/\ker(T)$ consisting of all cosets $[x] = x + \ker(T)$ and if one defines

$$\hat{T} : X/\ker(T) \to Y$$

by $\hat{T}(x+\ker(T)) = Tx$, then $\hat{T}$ is bijective. If one defines a norm on the quotient space $X/\ker(T)$ by

$$\|[x]\| = \inf\{\|x\| : x \in [x]\}$$

then $X/\ker(T)$ is a Banach space, $\hat{T}$ is continuous and $\|\hat{T}\| = \|T\|$ (proof?). Since $\hat{T}$ is also open

$$\hat{T}^{-1} : Y \to X/\ker(T)$$

is continuous and

$$\rho_T = \sup\{\rho > 0 : B[0;\rho] \subset TB[0;1]\} = (\|\hat{T}^{-1}\|)^{-1}$$

(proof?).

Now comes the main result of this section, the Theorem of

LYUSTERNIK or more precisely a generalization of the result obtained by LYUSTERNIK in 1934.

5.2.5 Theorem (LYUSTERNIK): Let X,Y be Banach spaces. $C \subset X$ and $K \subset Y$ nonempty, closed and convex. Let $g : X \to Y$ be a map which is continuously F-differentiable at the point

$$\bar{x} \in M := \{x \in X : x \in C,\ g(x) \in -K\}.$$

Moreover suppose

$$\text{(CQ)} \qquad g'(\bar{x})C(\bar{x}) + K(-g(\bar{x})) = Y$$

is satisfied. Then

$$\begin{aligned} L(M;\bar{x}) := \{h \in X : h \in C(\bar{x}),\ g'(\bar{x})h \in -K(-g(\bar{x}))\} \\ \subset \{h \in X : \exists\, t_o > 0 \text{ and } r : [0,t_o] \to X \text{ with} \\ \text{i) } \bar{x} + th + r(t) \in M \text{ for } t \in [0,t_o] \\ \text{ii) } \lim_{t\to 0+} \frac{r(t)}{t} = 0\} \\ \subset T(M;\bar{x}). \end{aligned}$$

Before we prove the theorem we wish to look once more at the special case $C = X$ and $K = \{0\}$. Then $M = \{x \in X : g(x) = 0\}$ and the condition (CQ) says that $g'(\bar{x}) \in L(X,Y)$ is surjective. The statement of the theorem is then that

$$\begin{aligned} \{h \in X : g'(\bar{x})h = 0\} \subset \{h \in X : \exists\, t_o > 0 \text{ and } r : [0,t_o] \to X \\ \text{with i) } g(\bar{x}+th+r(t)) = 0 \\ \text{for all } t \in [0,t_o] \text{ and} \\ \text{ii) } \lim_{t\to 0+} \frac{r(t)}{t} = 0\} \end{aligned}$$

and this is what LYUSTERNIK actually proved.

If $g : \mathbb{R}^n \to \mathbb{R}$, then $M = g^{-1}(0)$ is a hypersurface in $\mathbb{R}^n$ and $g'(\bar{x})$ is surjective if and only if $\nabla g(\bar{x}) \neq 0$. Every vector which is perpendicular to $\nabla g(\bar{x})$ is according to the theorem an element of the tangent cone, which is intuitively clear:

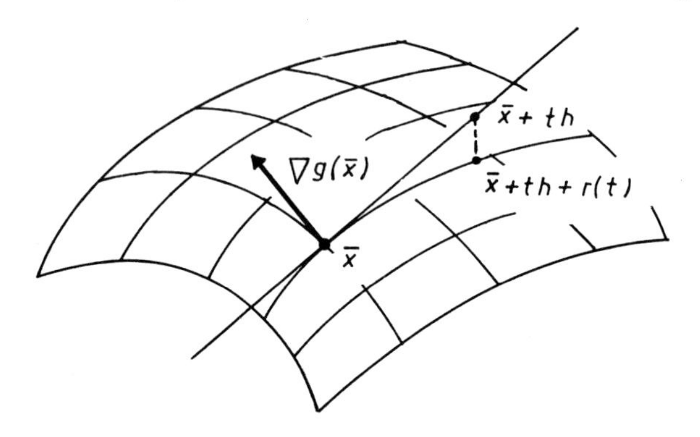

Proof of Theorem 5.2.5: 1) We first show:
to an arbitrarily given $h \in X$ there exist $\hat{t} > 0$, $c_0 > 0$ and maps $r : [0,\hat{t}] \to C(\bar{x})$, $z : [0,\hat{t}] \to K(-g(\bar{x}))$ with

a) $$\begin{Bmatrix} r(t) \\ z(t) \end{Bmatrix} \in c_0 \| g(\bar{x}+th)-g(\bar{x})-tg'(\bar{x})h \| \begin{Bmatrix} (C-\bar{x})_1 \\ (K+g(\bar{x}))_1 \end{Bmatrix} \text{ for } t \in [0,\hat{t}]$$

b) $$g(\bar{x}) + tg'(\bar{x})h = g(\bar{x}+th+r(t)) + z(t) \text{ for } t \in [0,\hat{t}].$$

To prove this we define as in Cor. 5.2.4:

$$\rho_0 := \sup \{\rho > 0 : B[0;\rho] \subset g'(\bar{x})(C-\bar{x})_1 + (K+g(\bar{x}))_1\}.$$

Let $\varepsilon \in (0,\frac{1}{2}\rho_0)$ be arbitrary. Since g is continuously F-differentiable at $\bar{x}$ there exists a $\delta > 0$ with

$$\| g'(x)-g'(\bar{x}) \| \leq \varepsilon \text{ for all } x \in B[\bar{x};2\delta].$$

By the generalized mean value theorem 5.1.3 ii) we have

$$\| g(x)-g(x')-g'(\bar{x})(x-x') \| \leq \varepsilon \| x-x' \|$$

for all $x,x' \in B[\bar{x};2\delta]$. Now choose a $\gamma > 1$ with $\gamma\left(\frac{1}{2} + \frac{\varepsilon}{\rho_0}\right) \leq 1$. Without loss of generality we may assume $h \neq 0$. Let

$\hat{t} := \delta / \| h \|$ and choose $t \in [0,\hat{t}]$ fixed.

We construct sequences $\{r_k\} \subset C(\bar{x})$, $\{z_k\} \subset K(-g(\bar{x}))$ as follows:

Set $r_o := 0$, $z_o := 0$.
Assume r_k and z_k have been chosen. Then by Cor. 5.2.4 there exist

$$\begin{Bmatrix} u_k \\ v_k \end{Bmatrix} \in \frac{\gamma}{\rho_o} \| g(\bar{x})+tg'(\bar{x})h-g(\bar{x}+th+r_k)-z_k \| \begin{Bmatrix} (C-\bar{x})_1 \\ (K+g(\bar{x}))_1 \end{Bmatrix}$$

with $g(\bar{x}) + tg'(\bar{x})h - g(\bar{x}+th+r_k) - z_k = g'(\bar{x})u_k + v_k$.
Set $r_{k+1} := r_k + u_k$, $z_{k+1} := z_k + v_k$.

We shall show that $\{r_k\}$ and $\{z_k\}$ are Cauchy sequences and that the limits $r = r(t)$ and $z = z(t)$ satisfy a) and b).

We make the following abbreviations:

$$d(t) := \| g(\bar{x}+th)-g(\bar{x})-tg'(\bar{x})h \|$$

(one has $d(t) \leq \varepsilon t \| h \| \leq \varepsilon\delta < \frac{\delta}{2}\rho_o$) and

$$q := \frac{\varepsilon\gamma}{\rho_o} \quad (\leq 1 - \tfrac{\gamma}{2} < \tfrac{1}{2}).$$

By induction we demonstrate that

i) $$\begin{Bmatrix} r_k \\ z_k \end{Bmatrix} \in \frac{\gamma}{\rho_o} d(t) \frac{1-q^k}{1-q} \begin{Bmatrix} (C-\bar{x})_1 \\ (K+g(\bar{x}))_1 \end{Bmatrix}$$

ii) $$\| g(\bar{x})+tg'(\bar{x})h-g(\bar{x}+th+r_k)-z_k \| \leq d(t)q^k$$

iii) $$\begin{Bmatrix} u_k \\ v_k \end{Bmatrix} \in \frac{\gamma}{\rho_o} d(t)q^k \begin{Bmatrix} (C-\bar{x})_1 \\ (K+g(\bar{x}))_1 \end{Bmatrix} .$$

(From i) it follows in particular that $\| r_k \| < \frac{\gamma}{\rho_o} \frac{\delta}{2} \rho_o \cdot \frac{1}{1-q} \leq \delta$, so that $\bar{x} + th + r_k \in B[\bar{x};2\delta]$.)

For $k = 0$ i), ii) and iii) are true. Suppose they also hold for k. Then we have

i) $\quad r_{k+1} = r_k + u_k$

$$\in \frac{\gamma}{\rho_o} d(t)\left(\frac{1-q^k}{1-q} + q^k\right)(C-\bar{x})_1 = \frac{\gamma}{\rho_o} d(t) \frac{1-q^{k+1}}{1-q} (C-\bar{x})_1$$

(use the induction hypotheses i), iii)). The statement for z_{k+1} follows similarly.

ii) $\quad \| g(\bar{x})+tg'(\bar{x})h-g(\bar{x}+th+r_{k+1})-z_{k+1} \| =$

$$\| g(\bar{x})+tg'(\bar{x})h-g(\bar{x}+th+r_k+u_k)-z_k-v_k \| =$$

$$\| g(\bar{x}+ th+r_k+u_k)-g(\bar{x}+th+r_k)-g'(\bar{x})u_k \| \le \varepsilon \| u_k \|$$

$$\le d(t)q^{k+1} \quad \text{(use the induction hypothesis iii)).}$$

iii) $\quad u_{k+1} \in \frac{\gamma}{\rho_o} \| g(\bar{x})+tg'(\bar{x})h-g(\bar{x}+th+r_{k+1})-z_{k+1} \| (C-\bar{x})_1$

$$\subset \frac{\gamma}{\rho_o} d(t)q^{k+1}(C-\bar{x})_1 \quad \text{(use ii)).}$$

The statement for v_{k+1} follows similarly.

Here we have used:

$$\alpha(C-\bar{x})_1 + \beta(C-\bar{x})_1 = (\alpha+\beta)(C-\bar{x})_1 \text{ for } \alpha,\beta \ge 0$$

and

$$\alpha(C-\bar{x})_1 \subset \beta(C-\bar{x})_1 \text{ for } 0 \le \alpha \le \beta.$$

i), ii), iii) are thus proved and it follows that

$$\{r_k\} \subset \frac{\gamma}{\rho_o} d(t) \frac{1}{1-q} (C-\bar{x})_1 \subset \frac{2}{\rho_o} d(t)(C-\bar{x})_1 \text{ and}$$

$$\{z_k\} \subset \frac{\gamma}{\rho_o} d(t) \frac{1}{1-q} (K+g(\bar{x}))_1 \subset \frac{2}{\rho_o} d(t)(K+g(\bar{x}))_1$$

are Cauchy sequences and hence converge to an

$$r = r(t) \in \frac{2}{\rho_o} d(t)(C-\bar{x})_1 \text{ resp.}$$

$$z = z(t) \in \frac{2}{\rho_o} d(t)(K+g(\bar{x}))_1.$$

By iii) $\{u_k\}$ and $\{v_k\}$ are sequences that converge to 0 and since

$$g(\bar{x}) + tg'(\bar{x})h - g(\bar{x}+th+r_k) - z_k = g'(\bar{x})u_k + v_k$$

it follows that

$$g(\bar{x}) + tg'(\bar{x})h = g(\bar{x}+th+r(t)) + z(t),$$

which proves 1) (set $c_o := 2/\rho_o$).

2) Suppose $h \in L(M;\bar{x}) = \{h \in X : h \in C(\bar{x}),\ g'(\bar{x})h \in -K(-g(\bar{x}))\}$. Let $r : [0,\hat{t}] \to C(\bar{x})$, $z : [0,\hat{t}] \to K(-g(\bar{x}))$ are chosen as in 1). Because

$$\frac{\| r(t) \|}{t} \le c_o \frac{1}{t} \| g(\bar{x}+th)-g(\bar{x})-tg'(\bar{x})h \|$$

we have

$$\lim_{t\to 0+} \frac{r(t)}{t} = 0.$$

It remains to show that $\bar{x} + th + r(t) \in M$ for all $t \in [0,t_o]$ with sufficiently small positive $t_o \le \hat{t}$.

We have $h \in C(\bar{x})$, i.e. $h = \lambda(c-\bar{x})$ with a $\lambda \ge 0$, $c \in C$. By 1) a)

$$r(t) = \lambda(t)(c(t)-\bar{x}) \text{ with } 0 \le \lambda(t),\ \lim_{t\to 0+} \lambda(t) = 0$$

and $c(t) \in C$. Since C is convex we have

$$\bar{x} + th + r(t) = (1-\lambda t-\lambda(t))\bar{x} + \lambda tc + \lambda(t)c(t) \in C,$$

provided $1 - \lambda t - \lambda(t) \ge 0$, which is the case for all sufficiently small $t > 0$.

Similarly from $g'(\bar{x})h \in -K(-g(\bar{x}))$ it follows that

$$g'(\bar{x})h = -\mu(k+g(\bar{x})) \text{ with a } \mu \geq 0,\ k \in K.$$

By 1) a) we have

$$z(t) = \mu(t)(k(t)+g(\bar{x})) \text{ with } 0 \leq \mu(t),\ \lim_{t\to 0+} \mu(t) = 0$$

and $k(t) \in K$. By 1) b) we have

$$g(\bar{x}+th+r(t)) = g(\bar{x}) + tg'(\bar{x})h - z(t)$$

$$= -((1-\mu t-\mu(t))(-g(\bar{x}) + \mu tk + \mu(t)k(t))$$

$$\in -K \text{ for all sufficiently small } t > 0.$$

Hence the theorem is completely proved.

<u>Remarks</u>: 1) The cone

$$A(M;\bar{x}) := \{h \in X : \exists\ t_o > 0 \text{ and } r : [0,t_o] \to X \text{ with}$$

$$\text{i)}\quad \bar{x} + th + r(t) \in M \text{ for } t \in [0,t_o]$$

$$\text{ii)}\quad \lim_{t\to 0+} \frac{r(t)}{t} = 0\}$$

is called the cone of attainable directions to M at $\bar{x}$ in e.g. BAZARAA-SHETTY [3, p. 81]. Under the hypotheses of the Theorem of LYUSTERNIK one has

$$L(M;\bar{x}) \subset A(M;\bar{x}) \subset T(M;\bar{x}).$$

2) The condition (CQ) $g'(\bar{x})C(\bar{x}) + K(-g(\bar{x})) = Y$ is called a regularity condition or constraint qualification. If e.g. $C = X$ (no explicit constraints) and $K = \{0\}$ (only equality constraints), then (CQ) means that $g'(\bar{x})X = Y$ resp. that $g'(\bar{x})$ is surjective. If in addition $X = \mathbb{R}^n$, $Y = \mathbb{R}^m$, i.e.

$$g(x) = (g_1(x),\dots,g_m(x))^T,$$

then $g'(\bar{x})$ is surjective if and only if $\nabla g_1(\bar{x}),\dots,\nabla g_m(\bar{x})$ are linearly independent (proof?).

In an important special case one can do without the completeness assumption on X in the Theorem of LYUSTERNIK. This we formulate as

5.2.6 Corollary: Let X,Y be normed linear spaces, Y finite dimensional. Let $g : X \to Y$ be continuously F-differentiable at $\bar{x} \in M := g^{-1}(0)$ and $g'(\bar{x})X = Y$. Then

$$L(M;\bar{x}) := \{h \in X : g'(\bar{x})h = 0\}$$

$$\subset \{h \in X : \exists\, t_o > 0 \text{ and } r : [0,t_o] \to X \text{ with}$$

$$\text{i)} \quad g(\bar{x}+th+r(t)) = 0 \text{ for } t \in [0,t_o]$$

$$\text{ii)} \quad \lim_{t\to 0+} \frac{r(t)}{t} = 0\} := A(M;\bar{x}).$$

Proof: Let dim Y = m and $Y = \text{span}\,\{y_1,\dots,y_m\}$. Since $g'(\bar{x})X = Y$ there exist $x_1,\dots,x_m \in X$ with $g'(\bar{x})x_i = y_i$ $(i=1,\dots,m)$. Then $x_1,\dots,x_m$ are also linearly independent; set

$$X_m := \text{span}\,\{x_1,\dots,x_m\}.$$

From $g'(\bar{x})X_m = Y$ it follows from 5.2.3 that

$$0 < \rho_o := \sup\,\{\rho > 0 : B[0;\rho] \subset g'(\bar{x})(X_m \cap B[0;1])\}.$$

Then one can take over the first part of the proof of Theorem 5.2.5 and gets for arbitrary $h \in X$ constants $\hat{t} > 0$, $c_o > 0$ and a map $r : [0,\hat{t}] \to X_m$ with

a) $\qquad \| r(t) \| \le c_o \| g(\bar{x}+th) - g(\bar{x}) - tg'(\bar{x})h \|$

b) $\qquad g(\bar{x}) + tg'(\bar{x})h = g(\bar{x}+th+r(t))$

for all $t \in [0,\hat{t}]$ and the claim follows.

5.3 LAGRANGE multipliers. Theorems of KUHN-TUCKER and F. JOHN type

The objective of this section is to set up and prove necessary optimality conditions of first order (i.e. only derivatives up to first order are used) for a local solution $\overline{x}$ of the optimization problem

$$\text{(P)} \qquad \text{Minimize } f(x) \text{ on } M := \{x \in X : x \in C,\ g(x) \in -K\}.$$

We remind the reader: $\overline{x} \in M$ is a <u>local</u> solution of (P) if there exists a neighborhood U of $\overline{x}$ with $f(\overline{x}) \leq f(x)$ for all $x \in M \cap U$. Convex programs (f convex, M convex), which we studied in the last chapter, are among other things distinguished by the fact that a local solution is also a global solution (proof?).

We further remind the reader of the notation $C(\overline{x})$ for the cone generated by $C - \overline{x}$ and of the Definition 4.2.1 of the cone dual to $A \subset X$:

$$A^+ := \{x^* \in X^* : \langle x^*, a\rangle \geq 0 \text{ for all } a \in A\}.$$

Since we always assume that $f : X \to R$ and $g : X \to Y$ are F-differentiable at $\overline{x}$, the following definition makes sense:

<u>5.3.1 Definition</u>: $\overline{y}^* \in Y^*$ is called a <u>Lagrange multiplier</u> for (P) at $\overline{x} \in M$ if

i) $\overline{y}^* \in K^+$

ii) $\langle \overline{y}^*, g(\overline{x})\rangle = 0$

iii) $f'(\overline{x}) + \overline{y}^* \circ g'(\overline{x}) \in C(\overline{x})^+$

$(\overline{x}, \overline{y}^*)$ is then called a KUHN-TUCKER pair and a theorem that guarantees the existence of a Lagrange multiplier for a local solution $\overline{x}$ of (P) is called a theorem of KUHN-TUCKER type or a Lagrange multiplier rule.

<u>Example</u>: Consider the problem of minimizing $f : \mathbb{R}^n \to R$ subject to the inequality constraints

$$g_1(x) \leq 0, \ldots, g_m(x) \leq 0$$

and the equality constraints

$$g_{m+1}(x) = 0, \ldots, g_{m+k}(x) = 0.$$

In the problem mentioned at the beginning then we must take $X = \mathbb{R}^n$, $Y = \mathbb{R}^{m+k}$, $C = X$ and $K = \{y = (y_i) \in \mathbb{R}^{m+k} : y_i \geq 0$ $(i=1,\ldots,m)$, $y_i = 0$ $(i=m+1,\ldots,m+k)\}$. Then

$$K^+ = \{y = (y_i) \in \mathbb{R}^{m+k} : y_i \geq 0 \ (i=1,\ldots,m)\}$$

and $C(\overline{x})^+ = \{0\}$. A vector $\overline{y} \in \mathbb{R}^{m+k}$ is thus a Lagrange multiplier at a feasible point $\overline{x}$ if and only if

i) $\qquad \overline{y}_i \geq 0 \ (i=1,\ldots,m)$

ii) $\qquad \overline{y}_i g_i(\overline{x}) = 0 \ (i=1,\ldots,m)$

iii) $\qquad \nabla f(\overline{x}) + \sum_{i=1}^{m+k} \overline{y}_i \nabla g_i(\overline{x}) = 0.$

If e.g. $n = 2$, $m = 3$ and $k = 0$, then the region of feasible points could look as follows:

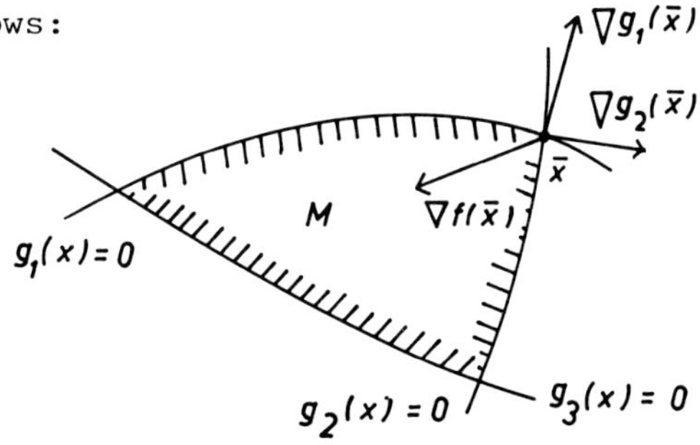

Suppose there is a local solution of the problem of minimizing f on M at $\overline{x} \in M$ with $g_1(\overline{x}) = g_2(\overline{x}) = 0$. Then it is intuitively clear that one can express $-\nabla f(\overline{x})$ as a nonnegative linear combination of $\nabla g_1(\overline{x})$ and $\nabla g_2(\overline{x})$ and this is exactly the assertion that a Lagrange multiplier exists.

The following theorem (c.f. ZOWE-KURCYUSZ [81]) is our first theorem of KUHN-TUCKER type.

5.3.2 Theorem: Suppose given the optimization problem

(P) Minimize $f(x)$ on $M := \{x \in X : x \in C,\ g(x) \in -K\}$.

Assume the following hold:

(V) i) $f : X \to \mathbb{R}$, X a Banach space.

ii) $C \subset X$ is nonempty, closed and convex.

iii) $g : X \to Y$, Y a Banach space.

iv) $K \subset Y$ is a nonempty, closed convex cone.

v) $\bar{x} \in M$ is a local solution of (P), f is F-differentiable and g continuously F-differentiable at $\bar{x}$.

Moreover assume the constraint qualification holds:

(CQ) $g'(\bar{x})C(\bar{x}) + K(-g(\bar{x})) = Y$.

Then there exists a Lagrange multiplier $\bar{y}^*$ for (P) at $\bar{x}$ (c.f. Definition 5.3.1).

Proof: The proof is in two parts:

1) $\bar{h} = 0$ is a solution of the linearized program

(PL) Minimize $f'(\bar{x})h$ on

$$L(M;\bar{x}) := \{h \in X : h \in C(\bar{x}), g'(\bar{x})h \in -K(-g(\bar{x}))\}.$$

2) For

$$\Lambda_o := \{(g'(\bar{x})h+z, f'(\bar{x})h+r) : h \in C(\bar{x}), z \in K(-g(\bar{x})), r \geq 0\}$$

one has $\operatorname{int}(\Lambda_o) \cap \{0\} \times \mathbb{R} \neq \emptyset$. Thus the constraint qualification of Theorem 4.3.2 is satisfied and the program dual to (PL)

(DL) Maximize $\varphi(y^*) := \inf\{f'(\bar{x})h + \langle y^*, g'(\bar{x})h\rangle : h \in C(\bar{x})\}$

$$\text{on } N := \{y^* \in Y^* : y^* \in K(-g(\bar{x}))^+, \varphi(y^*) > -\infty\}$$

has a solution $\overline{y}^*$ and it is the desired Lagrange multiplier.

1) Let $h \in L(M;\overline{x})$ be arbitrary. The LYUSTERNIK Theorem 5.2.5 gives the existence of a $t_o > 0$ and a map $r : [0,t_o] \to X$ with $\overline{x} + th + r(t) \in M$ for $t \in [0,t_o]$ and

$$\lim_{t\to 0+} \frac{r(t)}{t} = 0.$$

Since $\overline{x}$ is a local solution of (P) there exists a $\hat{t}_o \in (0,t_o]$ with

$$f(\overline{x}) \le f(\overline{x}+th+r(t)) \text{ for all } t \in [0,\hat{t}_o].$$

Thus

$$f'(\overline{x})h = \lim_{t\to 0+} \frac{1}{t} (f(\overline{x}+th+r(t))-f(\overline{x})) \ge 0.$$

2) We show that $(0,r_o) \in \text{int}\,(\Lambda_o)$ for $r_o > 0$. By the generalized open mapping theorem 5.2.3 there exists a $\rho > 0$ with

$$B[0;\rho] \subset g'(\overline{x})(C-\overline{x})_1 + (K+g(\overline{x}))_1.$$

If one now chooses $\varepsilon > 0$ so small that $\frac{\varepsilon}{\rho} \| f'(\overline{x}) \| \le \frac{1}{2} r_o$, then

$$(0,r_o) + B[0;\varepsilon] \times [-r_o/2, r_o/2] \subset \Lambda_o,$$

and hence $\text{int}\,(\Lambda_o) \cap \{0\} \times \mathbb{R} \neq \emptyset$. The program (DL) dual to (PL) thus has a solution $\overline{y}^*$ by Theorem 4.3.2 and we have

$$\max\,(DL) = \varphi(\overline{y}^*) = 0 = \min\,(PL).$$

From $\overline{y}^* \in K(-g(\overline{x}))^+$ one gets $\langle\overline{y}^*, z+\lambda g(\overline{x})\rangle \ge 0$ for all $z \in K$, $\lambda \ge 0$ and hence $\overline{y}^* \in K^+$ and $\langle\overline{y}^*, g(\overline{x})\rangle = 0$ (first set $\lambda = 0$ and then $z = 0$). From $\varphi(\overline{y}^*) > -\infty$ resp. the fact that

$$f'(\overline{x}) + \overline{y}^* \circ g'(\overline{x})$$

is bounded below on the cone $C(\overline{x})$ it follows that

$$f'(\bar{x}) + \bar{y}^* \circ g'(\bar{x}) \in C(\bar{x})^+.$$

Hence $\bar{y}^*$ is a Lagrange multiplier for (P) at $\bar{x}$ and the theorem is proved.

Remark: If in Theorem 5.3.2 $C = X$, i.e. if no explicit constraints occur, and if further Y is finite dimensional and $K = \{0\}$, i.e. if one only has finitely many equations as constraints, then X need only be a normed linear space (not necessarily complete). (Hint: Remember Corollary 5.2.6).

If in the optimization problem (P) the implicit constraint $g(x) \in - K$ is of the inequality type, i.e. if $\text{int}\,(K) \neq \emptyset$, then one can weaken the smoothness conditions on f,g and prove a theorem of KUHN-TUCKER type without the completeness of X,Y:

5.3.3 Theorem: Suppose given the optimization problem

(P) Minimize $f(x)$ on $M := \{x \in X : x \in C,\ g(x) \in - K\}$.

Suppose further: X,Y are normed linear spaces, $C \subset X$ nonempty and convex, $K \subset Y$ a convex cone with $\text{int}\,(K) \neq \emptyset$. Suppose $f : X \to \mathbb{R}$ and $g : X \to Y$ are G-differentiable at the local solution $\bar{x} \in M$. Finally suppose

$$L_o := \{h \in X : h \in C(\bar{x}), g(\bar{x}) + g'(\bar{x})h \in -\text{int}(K)\} \neq \emptyset.$$

Then there exists a Lagrange multiplier for (P) at $\bar{x}$.

Proof: The structure of the proof is analogous to that of Theorem 5.3.2.

1) For $L(M;\bar{x}) := \{h \in \bar{X} : h \in C(\bar{x}),\ g'(\bar{x})h \in - K(-g(\bar{x}))\}$ we have that $\bar{h} = 0$ is a solution of

(PL) Minimize $f'(\bar{x})h$ on $L(M;\bar{x})$.

2) Let

$$\Lambda_o := \{(g'(\bar{x})h+z, f'(\bar{x})h+r) : h \in C(\bar{x}), z \in K(-g(\bar{x})), r \geq 0\}.$$

Then int $(\Lambda_o) \cap \{0\} \times \mathbb{R} \neq \emptyset$, i.e. the program dual to (PL) has a solution $\bar{y}^*$ by Theorem 4.3.2 and it is a Lagrange multiplier for (P) at $\bar{x}$.

1) One defines the cone $F(M;\bar{x})$ of feasible directions to M at $\bar{x}$ by

$$F(M;\bar{x}) := \{h \in X : \exists\ t_o > 0 \text{ with } \bar{x} + th \in M \text{ for } t \in [0,t_o]\}$$

and shows a) $L_o \subset F(M;\bar{x})$

b) $L(M;\bar{x}) \subset \text{cl}\ (F(M;\bar{x}))$.

Since f is G-differentiable at the local solution $\bar{x}$ of (P) we have $f'(\bar{x})h \geq 0$ for all $h \in F(M;\bar{x})$, so 1) follows from b).

a) Let $h \in L_o$ be arbitrary, i.e. $h = \lambda(c-\bar{x})$ with $\lambda \geq 0$, $c \in C$ and

$$g(\bar{x}) + g'(\bar{x})h = g(\bar{x}) + \lim_{t \to 0+} \frac{1}{t}\,(g(\bar{x}+th)-g(\bar{x})) \in -\ \text{int}\ (K).$$

Then one can choose a $t_o \in (0,1]$ so small that $\lambda t_o \leq 1$ and

$$g(\bar{x}) + \frac{1}{t}\,(g(\bar{x}+th)-g(\bar{x})) \in -\ K \text{ for } t \in (0,t_o].$$

Then we have

$$\bar{x} + th = \bar{x} + \lambda t(c-\bar{x}) \in C$$

$$g(\bar{x}+th) = (1-t)g(\bar{x}) + t(g(\bar{x}) + \frac{1}{t}(g(\bar{x}+th)-g(\bar{x}))) \in -\ K$$

for $t \in (0,t_o]$, i.e. $h \in F(M;\bar{x})$.

b) Suppose $h \in L(M;\bar{x})$, i.e. $h \in C(\bar{x})$ and $g'(\bar{x})h = -\ \mu(z+g(\bar{x}))$ with $\mu \geq 0$, $z \in K$. Choose $h_o \in L_o$.

i) $\mu \in [0,1]$. Then $(1-t)h + th_o \in L_o$ for $t \in (0,1]$, since

$$g(\bar{x}) + g'(\bar{x})((1-t)h+th_o) =$$

$$t(g(\bar{x})+g'(\bar{x})h_o) + (1-t)((1-\mu)g(\bar{x})-\mu z) \in -\text{ int }(K).$$

By a) we have

$$h = \lim_{t\to 0+} ((1-t)h+th_o) \in \text{cl }(L_o) \subset \text{cl}(F(M;\bar{x})).$$

ii) $\mu > 1$. By i) we have $\frac{h}{\mu} \in \text{cl }(F(M;\bar{x}))$. Since $F(M;\bar{x})$ is a cone, so is $\text{cl}(F(M;\bar{x}))$, so b) is proved.

2) Since $L_o \neq \emptyset$ there exists an $\hat{h} \in C(\bar{x})$ with

$$g'(\bar{x})\hat{h} \in -\text{ int}(K) - g(\bar{x}) \subset -\text{ int }(K(-g(\bar{x}))).$$

Thus SLATER's constraint qualification is fulfilled for the program

(PL) Minimize $f'(\bar{x})h$ on $L(M;\bar{x})$

and a solution $\bar{y}^*$ of the program dual to (PL) is a Lagrange multiplier for (P) at $\bar{x}$.

<u>Remarks</u>: 1. The proofs of the theorems 5.3.2 and 5.3.3 were so formulated as to make it clear that the desired Lagrange multiplier is found by linearizing the nonlinear program (P) at the local solution $\bar{x}$ and then solving the program dual to this linearized program.

2. In the proof of Theorem 5.3.3 we saw that the regularity condition

$$L_o := \{h \in X : h \in C(\bar{x}), g(\bar{x}) + g'(\bar{x})h \in -\text{ int }(K)\} \neq \emptyset$$

implies that the SLATER constraint qualification holds for the linearized program (PL). Furthermore:

i) $L_o \neq \emptyset \Rightarrow \exists\ \hat{x} \in C$ with $g(\hat{x}) \in -\text{ int }(K)$ (set $\hat{x} = \bar{x} + t_o h_o$ with $h_o \in L_o$ and sufficiently small $t_o > 0$).

ii) If $g : X \to Y$ is K-convex (c.f. Definition 3.3.11) and the convex cone $K \subset Y$ is closed then:
$\exists\ \hat{x} \in C$ with $g(\hat{x}) \in -\operatorname{int}(K) \Rightarrow L_o \neq \emptyset$. For: $h := \hat{x} - \bar{x} \in L_o$, since

$$g(\hat{x}) - g(\bar{x}) - g'(\bar{x})(\hat{x}-\bar{x}) = \lim_{t\to 0+} \frac{(1-t)g(\bar{x})+tg(\hat{x})-g(\bar{x}+t(\hat{x}-\bar{x}))}{t}$$

$$\in \operatorname{cl}(K) = K$$

and thus $g(\bar{x}) + g'(\bar{x})(\hat{x}-\bar{x}) \in g(\hat{x}) - K \subset -\operatorname{int}(K)$.

3. One readily shows that:

$$L_o := \{h \in X : h \in C(\bar{x}), g(\bar{x}) + g'(\bar{x})h \in -\operatorname{int}(K)\} \neq \emptyset$$

$$\Rightarrow g'(\bar{x})C(\bar{x}) + K(-g(\bar{x})) = Y.$$

With stronger hypotheses (completeness of X,Y, closure of C,K etc.) one could also have obtained the conclusion of Theorem [illegible].3 from Theorem 5.3.2.

4. Without some constraint qualification

$$\text{(CQ)} \qquad g'(\bar{x})C(\bar{x}) + K(-g(\bar{x})) = Y$$

resp. $L_o \neq \emptyset$ one cannot expect to have the existence of a Lagrange multiplier. Here is an example (c.f. LUENBERGER [53, p. 249]):Consider the problem

$$\text{Minimize } f(x_1,x_2) = -x_1$$

$$\text{on } M := \{(x_1,x_2) \in \mathbb{R}^2 : g(x_1,x_2) = \begin{pmatrix} -x_1 \\ -x_2 \\ x_2-(1-x_1)^3 \end{pmatrix} \leq 0\}.$$

$\bar{x} = \begin{pmatrix} 1 \\ 0 \end{pmatrix}$ is a solution for which there is no Lagrange multiplier:

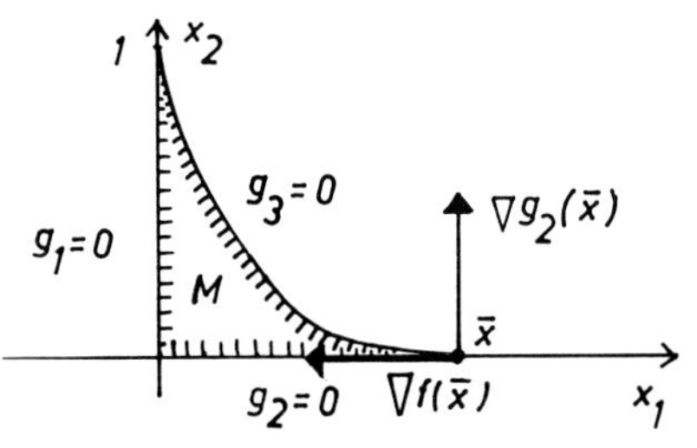

$\nabla g_2(\bar{x}) = -\nabla g_3(\bar{x})$, $\nabla f(\bar{x})$ is linearly independent of $\nabla g_2(\bar{x})$. Thus no Lagrange multiplier can exist.

In one very important special case no constraint qualification is needed in order to prove a theorem of KUHN-TUCKER type - namely for finite dimensional optimization problems for which all constraints are affine linear.

5.3.4 Theorem: Suppose given the finite dimensional optimization problem with linear constraints

(P) Minimize $f(x)$ on

$$M := \{x \in \mathbb{R}^n : a^{iT}x - \beta_i \leq 0 \ (i=1,\ldots,m),$$
$$a^{iT}x - \beta_i = 0 \ (i=m+1,\ldots,m+k)\}.$$

Let $\bar{x} \in M$ be a local solution of (P), $f : \mathbb{R}^n \to \mathbb{R}$ continuously partially differentiable at $\bar{x}$. Then there exists a $\bar{y} = (\bar{y}_i) \in \mathbb{R}^{m+k}$ with

i) $\bar{y}_i \geq 0 \ (i=1,\ldots,m)$

ii) $\bar{y}_i(a^{iT}\bar{x}-\beta_i) = 0 \ (i=1,\ldots,m)$

iii) $\nabla f(\bar{x}) + \sum_{i=1}^{m+k} \bar{y}_i \ a^i = 0.$

i.e. there exists a Lagrange multiplier for (P) at $\bar{x}$.

Proof: 1) Let

$$I(\bar{x}) := \{i \in \{1,\ldots,m\} : a^{iT}\bar{x} - \beta_i = 0\}.$$

Then

$$F(M;\bar{x}) := \{h \in \mathbb{R}^n : \exists\ t_o > 0 \text{ with } \bar{x} + th \in M \text{ for all } t \in [0,t_o]\}$$

$$= \{h \in \mathbb{R}^n : a^{iT}h \le 0 \ (i \in I(\bar{x}))$$

$$a^{iT}h = 0 \ (i=m+1,\dots,m+k)\}.$$

Since $\bar{x}$ is a local solution of (P), $\bar{h} = 0$ is a solution of the linear program

(PL) Minimize $\nabla f(\bar{x})^T h$ on $F(M;\bar{x})$.

2) From the duality theory in linear optimization it follows that the program dual to (PL) has a solution, in particular a feasible solution, and a simple calculation proves the claim.

One should note that we have already proved the last theorem for a special case (c.f. Theorem 4.4.2). Furthermore we should emphasize once again that the existence of a Lagrange multiplier at $\bar{x}$ for convex programs (P) (f convex), g convex with respect to the closed convex cone $K \subset Y$) is also sufficient for $\bar{x} \in M$ to be a solution of (P) (proof?).

Now we must take a closer look at the constraint qualification

(CQ) $g'(\bar{x})C(\bar{x}) + K(-g(\bar{x})) = Y$

and consider the case that among the implicit constraints $g(x) \in -K$ there are both inequality and equality conditions.

5.3.5 Lemma: Suppose X, Y_1, Y_2 are normed linear spaces, $C \subset X$ is convex, $K_1 \subset Y_1$ a convex cone with cor $(K_1) \neq \emptyset$. Let

$$g : X \to Y := Y_1 \times Y_2, g(x) = (g_1(x), g_2(x))$$

be G-differentiable at

$$\bar{x} \in M := \{x \in X : x \in C, g(x) \in -K := -K_1 \times \{0\}\}.$$

Then the condition

(CQ) $\quad g'(\bar{x})C(\bar{x}) + K(-g(\bar{x})) = Y$

is satisfied if

i) cor (C) $\neq \emptyset$ and
$\{c_o \in \text{cor}(C) : g_1(\bar{x}) + g_1'(\bar{x})(c_o-\bar{x}) \in -\text{cor}(K_1),$
$g_2'(\bar{x})(c_o-\bar{x}) = 0\} \neq \emptyset;$

ii) $g_2'(\bar{x})X = Y_2$

Proof: Let $(y_1,y_2) \in Y_1 \times Y_2$ be arbitrarily given. By i) there exists a $c_o \in \text{cor}(C)$ with $k_o := -g_1(\bar{x}) - g_1'(\bar{x})(c_o-\bar{x}) \in \text{cor}(K_1)$, $g_2'(\bar{x})(c_o-\bar{x}) = 0$. By ii) there exists an $x \in X$ with $g_2'(\bar{x})x = y_2$. Since $c_o \in \text{cor}(C)$ it follows

$$c := c_o + \frac{1}{\lambda} x \in C$$

for sufficiently large $\lambda > 0$ and thus

$$h := x + \lambda(c_o-\bar{x}) = \lambda(c-\bar{x}) \in C(\bar{x})$$

$$g_2'(\bar{x})h = y_2.$$

Then one chooses a $\mu > 0$ so large that

$$k := k_o + \frac{1}{\mu}(y_1-g_1'(\bar{x})h) \in K_1.$$

Then

$$h + \mu(c_o-\bar{x}) = \lambda(c-\bar{x}) + \mu(c_o-\bar{x}) \in C(\bar{x}),$$

$$g_1'(\bar{x})(h+\mu(c_o-\bar{x})) + \mu(k+g_1(\bar{x})) = y_1 \text{ and}$$

$$g_2'(\bar{x})(h+\mu(c_o-\bar{x})) = y_2,$$

i.e. $(y_1,y_2) \in g'(\bar{x})C(\bar{x}) + K(-g(\bar{x}))$, which proves (CQ).

Example: If in Lemma 5.3.4 $X = \mathbb{R}^n$, $Y_1 = \mathbb{R}^m$ and $K_1 \subset Y_1$ is the nonnegative orthant, $Y_2 = \mathbb{R}^k$ and for simplicity $C = X$ and further

$$M := \{x \in \mathbb{R}^n : g_i(x) \leq 0 \ (i=1,\ldots,m),$$

$$g_i(x) = 0 \ (i=m+1,\ldots,m+k)\},$$

then the constraint qualification (CQ) is fulfilled at any point $\bar{x} \in M$ at which the g_i are continuously partially differentiable provided

i) $\exists\, h \in \mathbb{R}^n$ with $\nabla g_i(\bar{x})^T h < 0$ for all $i \in \{1,\ldots,m\}$ with with $g_i(\bar{x}) = 0$

$$\nabla g_i(\bar{x})^T h = 0 \text{ for all } i = m+1,\ldots,m+k.$$

ii) $\nabla g_{m+1}(\bar{x}),\ldots,\nabla g_{m+k}(\bar{x})$ are linearly independent.

(This is the so-called ARROW-HURWICZ-UZAWA constraint qualification, c.f. e.g. MANGASARIAN [57, p. 172].) If in addition $\bar{x} \in M$ is a local solution of the problem of minimizing $f(x)$ on M and $f : \mathbb{R}^n \to \mathbb{R}$ is continuously partially differentiable at $\bar{x}$, then by Theorem 5.3.2 there exists a Lagrange multiplier $\bar{y} \in \mathbb{R}^{m+k}$ (c.f. the example following Definition 5.3.1).

Without a constraint qualification one cannot expect the existence of a Lagrange multiplier. One can however still hope for a theorem of F. JOHN type. Instead of explaining what we mean by that we simply give such a theorem.

<u>5.3.6 Theorem:</u> Suppose given the optimization problem

(P) Minimize $f(x)$ on $M := \{x \in X : x \in C, g(x) \in -K\}$.

Suppose the hypotheses (V) i) - v) of Theorem 5.3.2 are fulfilled. Further suppose that $g'(\bar{x})C(\bar{x}) + K(-g(\bar{x}))$ is closed in Y or that Y is finite dimensional. Then there exists a

$$(\bar{y}_o,\bar{y}^*) \in \mathbb{R} \times Y^* \setminus \{(0,0)\}$$

with

i) $\bar{y}_o \geq 0,\ \bar{y}^* \in K^+$

ii) $\langle \bar{y}^*, g(\bar{x}) \rangle = 0$

iii) $\bar{y}_o f'(\bar{x}) + \bar{y}^* \circ g'(\bar{x}) \in C(\bar{x})^+$.

Proof: 1) Suppose $g'(\bar{x})C(\bar{x}) + K(-g(\bar{x}))$ is closed in Y. Without loss of generality we may assume that $g'(\bar{x})C(\bar{x}) + K(-g(\bar{x})) \subsetneqq Y$, since otherwise one can apply Theorem 5.3.2 and the claim follows with $\bar{y}_o = 1$. Then one can use the strict separation theorem to separate a point $y \in Y$ not belonging to the closed convex set $g'(\bar{x})C(\bar{x}) + K(-g(\bar{x}))$ strictly from the latter. Thus there exists a $\bar{y}^* \in Y^* \setminus \{0\}$ with

$$\langle \bar{y}^*, y \rangle < \langle \bar{y}^*, g'(\bar{x})h+z \rangle$$

for all $h \in C(\bar{x})$, $z \in K(-g(\bar{x}))$. By standard arguments one gets $\bar{y}^* \in K^+$, $\langle \bar{y}^*, g(\bar{x}) \rangle = 0$ and $\bar{y}^* \circ g'(\bar{x}) \in C(\bar{x})^+$. Then with $\bar{y}_o = 0$ the claim of the theorem is proved.

2) Let Y be finite dimensional. In 1) we showed that if

$$g'(\bar{x})C(\bar{x}) + K(-g(\bar{x})) = Y \text{ resp. } \mathrm{cl}(g'(\bar{x})C(\bar{x}) + K(-g(\bar{x}))) \subsetneqq Y$$

then the conclusion of the theorem is true for $\bar{y}_o = 1$ resp. $\bar{y}_o = 0$. But if Y is finite dimensional then one of these cases must hold. For a convex proper subset of a finite dimensional space cannot be dense, i.e. the closure is still proper (proof?).

Remark: For infinite dimensional Y one can have

$$g'(\bar{x})C(\bar{x}) + K(-g(\bar{x})) \subsetneqq \mathrm{cl}(g'(\bar{x})C(\bar{x})+K(-g(\bar{x}))) = Y$$

and then a nontrivial pair $(\bar{y}_o, \bar{y}^*) \in \mathbb{R} \times Y^*$ with the properties i), ii), iii) of Theorem 5.3.6 need not necessarily exist. One finds an example in TICHOMIROV [74, p. 65].

We close this section with a variant of Theorem 5.3.6 which coincides with the latter if Y if finite dimensional:

<u>5.3.7 Theorem:</u> Suppose given the optimization problem

(P) Minimize $f(x)$ on $M := \{x \in X : x \in C,\ g(x) \in -K\}$.

Suppose the hypotheses (V) i) - v) of Theorem 5.3.2 are satisfied. Further suppose that $\operatorname{icr}(g'(\bar{x})C(\bar{x}) + K(-g(\bar{x}))) \neq \emptyset$. Then there exists a $(\bar{y}_o, \bar{y}') \in \mathbb{R} \times Y' \setminus \{(0,0)\}$ with

i) $\bar{y}_o \geq 0$, $\langle \bar{y}', k \rangle \geq 0$ for all $k \in K$

ii) $\langle \bar{y}', g(\bar{x}) \rangle = 0$

iii) $\langle \bar{y}_o f'(\bar{x}) + \bar{y}' \circ g'(\bar{x}), h \rangle \geq 0$ for all $h \in C(\bar{x})$.

<u>Proof:</u> Apply Theorem 5.3.2 $(g'(\bar{x})C(\bar{x})+K(-g(\bar{x})) = Y)$ resp. Theorem 3.1.12 $(g'(\bar{x})C(\bar{x})+K(-g(\bar{x})) \subsetneq Y)$.

<u>Remark:</u> If we consider the problem

(P_o) Minimize $f(x)$ on $M_o := M \cap X_o$,

where $X_o \subset X$ is open, which is seemingly more general than

(P) Minimize $f(x)$ on $M := \{x \in X : x \in C,\ g(x) \in -K\}$,

then a local solution $\bar{x}$ of (P_o) is also a local solution of (P), so that for (P_o) all the necessary optimality conditions of this section hold as well.

5.4 <u>Necessary optimality conditions of first order in the calculus of variations and in optimal control theory</u>

Now we shall apply the results of the last section to some problems in the calculus of variations and optimal control theory. We shall proceed step by step starting with the simplest problems in the calculus of variations and finishing with a problem in optimal control theory. The following exposition has the character of a collection of examples; we can only try to give an impression -certainly not a complete treatment- of the topic of necessary optimality conditions in the calculus of variations and optimal control theory.

To facilitate comprehension we shall always precede each application by a functional analytic "model theorem".

1. The fixed end point problem.

Let X be a normed linear space, $C \subset X$ nonempty and convex, $X_o \subset X$ open and $f : X \to \mathbb{R}$. If $\bar{x} \in C \cap X_o$ is a local solution of

$$\text{Minimize } f(x) \text{ on } C \cap X_o$$

and f is G-differentiable at $\bar{x}$, then

$$f'(\bar{x})(c-\bar{x}) \geq 0 \text{ for all } c \in C.$$

5.4.1 Theorem: Let $X := C^1_{pc,n}[t_o,t_1]$ (piecewise continuously differentiable functions from $[t_o,t_1]$ to $\mathbb{R}^n$ with

$$\| x \| = \max (\| x \|_\infty, \| \dot{x} \|_\infty)$$

as norm). Suppose given the problem

$$\text{Minimize } I(x) = \int_{t_o}^{t_1} f(x(t),\dot{x}(t),t)dt \text{ on } C \cap X_o$$

(P) $\qquad$ with $C := \{x \in X : x(t_o) = x_o, x(t_1) = x_1\}$ and

$$X_o := \{x \in X : (x(t),\dot{x}(t)) \in U \text{ for all } t \in [t_o,t_1]\}.$$

We assume $x_o, x_1 \in \mathbb{R}^n$, $f : \mathbb{R}^n \times \mathbb{R}^n \times [t_o,t_1] \to \mathbb{R}$. Let $U \subset \mathbb{R}^n \times \mathbb{R}^n$ be open, f continuous on $U \times [t_o,t_1]$ and continuously partially differentiable with respect to $x,\dot{x}$. Let $\bar{x} \in C \cap X_o$ be a local solution of (P). Then there exists a $c \in \mathbb{R}^n$ with

$$\nabla_{\dot{x}} f(t) = \int_{t_o}^{t} \nabla_x f(s)ds + c \text{ for all } t \in [t_o,t_1]$$

(the EULER equation in integrated form), where we abbreviate

$$\nabla_x f(t) := \nabla_x f(\bar{x}(t), \dot{\bar{x}}(t), t)$$

and $\nabla_{\dot{x}} f(t)$ correspondingly.

If $\bar{x} \in C_n^1[t_o, t_1]$, then

$$\nabla_x f(t) - \frac{d}{dt} \nabla_{\dot{x}} f(t) = 0 \text{ for all } t \in [t_o, t_1]$$

(the EULER equation).

Proof: Since by our hypotheses I is even F-differentiable at $\bar{x}$ (c.f. Example 2) in 5.1) with

$$I'(\bar{x})h = \int_{t_o}^{t_1} \{\nabla_x f(t)^T h(t) + \nabla_{\dot{x}} f(t)^T \dot{h}(t)\} dt$$

and $C - \bar{x} = \{h \in X : h(t_o) = 0, h(t_1) = 0\} =: H$ is a linear subspace of X, we necessarily have

$$I'(\bar{x})h = 0 \text{ for all } h \in H.$$

By partial integration we get

$$\int_{t_o}^{t_1} \{-F(t) + \nabla_{\dot{x}} f(t)\}^T \dot{h}(t) dt = 0 \text{ for all } h \in H,$$

where $F(t) := \int_{t_o}^{t} \nabla_x f(s) ds$. Set

$$c := \frac{1}{t_1 - t_o} \int_{t_o}^{t_1} \{-F(t) + \nabla_{\dot{x}} f(t)\} dt \text{ and}$$

$$h(t) := \int_{t_o}^{t} \{-F(s) + \nabla_{\dot{x}} f(s) - c\} ds.$$

Then $h \in H$ and

$$\int_{t_o}^{t_1} |-F(t) + \nabla_{\dot{x}} f(t) - c|^2 dt = \int_{t_o}^{t_1} \{-F(t) + \nabla_{\dot{x}} f(t) - c\}^T \dot{h}(t) dt$$

$$= \int_{t_o}^{t_1} \{-F(t)+\nabla_{\dot{x}} f(t)\}^T \dot{h}(t)dt - c(h(t_1)-h(t_o))$$

$$= 0,$$

from which the claim follows.

Example: We return to Example 1) in 1.2, the BERNOULLI problem of the brachystochrone. Here one must minimize the functional

$$I(x) = \int_{t_o}^{t_1} \left(\frac{1+\dot{x}^2(t)}{x_o-x(t)}\right)^{1/2} dt$$

subject to the conditions $x(t_o) = x_o$, $x(t_1) = x_1$ $(x_1 < x_o)$.

Let us proceed absolutely unscrupulously and apply the EULER equation in spite of the singularity of the integrand

$$f(x,\dot{x}) = \left(\frac{1+\dot{x}^2}{x_o-x}\right)^{1/2} :$$

a solution $\bar{x}$ satisfies

$$f_x(\bar{x}(t),\dot{\bar{x}}(t)) - \frac{d}{dt} f_{\dot{x}}(\bar{x}(t),\dot{\bar{x}}(t)) = 0$$

(where f_x is the partial derivative of f with respect to x and correspondingly for $f_{\dot{x}}$). Since f does not depend on t, we have

$$\frac{d}{dt} \{f(\bar{x}(t),\dot{\bar{x}}(t)) - \dot{\bar{x}}(t) f_{\dot{x}}(\bar{x}(t),\dot{\bar{x}}(t))\} = 0$$

and thus by an easy calculation

$$(x_o-\bar{x}(t))(1+\dot{\bar{x}}^2(t)) = A, \; A = \text{const.}$$

Since $t_o < t_1$, $x_1 < x_o$ the slope of $\bar{x}$ will be negative so $\bar{x}$ is a solution of

$$\dot{x} = -\left(\frac{A-(x_o-x)}{x_o-x}\right)^{1/2}, \; x(t_o) = x_o.$$

This is a differential equation with separated variables, so

$$\int_{x}^{x_o} \left(\frac{x_o - x}{A-(x_o-x)}\right)^{1/2} dx = t - t_o .$$

If one substitutes

$$\tan^2 \frac{\varphi}{2} = \frac{x_o - x}{A-(x_o-x)} \quad \text{resp. } x = x_o - \frac{A}{2}\,(1-\cos\varphi)$$

then $dx = -\frac{A}{2}\sin\varphi\, d\varphi$ and thus

$$t - t_o = \frac{A}{2}\int_0^{\varphi} \tan\frac{\varphi}{2}\sin\varphi\, d\varphi = \frac{A}{2}\int_0^{\varphi}(1-\cos\varphi)d\varphi$$

$$= \frac{A}{2}\,(\varphi - \sin\varphi).$$

Hence for $\bar{x}$ one has the parametric presentation of a cycloid

$$\left.\begin{aligned} t &= t_o + \frac{A}{2}\,(\varphi - \sin\varphi) \\ \bar{x} &= x_o - \frac{A}{2}\,(1-\cos\varphi) \end{aligned}\right. \qquad 0 \le \varphi \le \varphi_1 .$$

where the still undetermined constants A, φ_1 can be computed from $\bar{x}(t_1) = x_1$ resp.

$$t_1 = t_o + \frac{A}{2}\,(\varphi_1 - \sin\varphi_1)$$

$$x_1 = x_o - \frac{A}{2}\,(1-\cos\varphi_1).$$

2. Variable end point problems

The shortest path between two points in the plane is a straight line segment. The shortest path between a point and a given smooth curve in the plane is a straight line segment which intersects the curve in a right angle. One calls this a

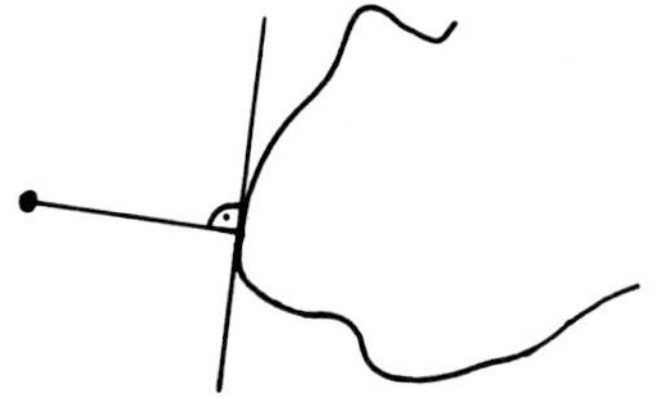

transversality condition. This condition joins the necessary optimality conditions as compensation for the fact that one of the endpoints is not a priori known.

We consider the problem

$$\text{Minimize } I(x,t_1) := \int_{t_o}^{t_1} f(x(t),\dot{x}(t),t)dt \text{ subject to}$$

$$x(t_o) = x_o,\ G(x(t_1),t_1) = 0$$

and wish to apply the following functional analytic "model theorem" (Theorem 5.3.6 resp. Theorem 5.3.2):

Let X be a normed linear space, $C \subset X$ convex and closed, $X_o \subset X$ open. Suppose given $f : X_o \to \mathbb{R}$ and $g : X_o \to \mathbb{R}^k$, $g(x) = (g_1(x),\ldots,g_k(x))^T$. If $\bar{x} \in C \cap g^{-1}(0) \cap X_o$ is a local solution of

$$\text{Minimize } f(x) \text{ on } C \cap g^{-1}(0) \cap X_o,$$

and if f is F-differentiable, g continuously F-differentiable at $\bar{x}$, then there exist $\bar{y}_o \geq 0$, $\bar{y} \in \mathbb{R}^k$ with $(\bar{y}_o,\bar{y}) \neq (0,0)$ and

$$(\bar{y}_o f'(\bar{x}) + \sum_{i=1}^{k} \bar{y}_i g_i'(\bar{x}))(c-\bar{x}) \geq 0 \text{ for all } c \in C.$$

If $g'(\bar{x})C(\bar{x}) = \mathbb{R}^k$, then one can choose $\bar{y}_o = 1$.

5.4.2 Theorem: Suppose $t_o < \hat{t}_1$ and $X := C^1_{pc,n}[t_o,\hat{t}_1] \times \mathbb{R}$. Suppose given the problem

(P)

$$\text{Minimize } I(x,t_1) := \int_{t_o}^{t_1} f(x(t),\dot{x}(t),t)dt \text{ on } C \cap g^{-1}(0) \cap X_o$$

with $C := \{(x,t_1) \in X : x(t_o) = x_o\}$

$$X_o := \{(x,t_1) \in X : t_1 \in (t_o,\hat{t}_1)\}$$

and $g : X_o \to \mathbb{R}^k$ defined by $g(x,t_1) = G(x(t_1),t_1)$.

Suppose $x_o \in \mathbb{R}^n$, and that $f : \mathbb{R}^n \times \mathbb{R}^n \times [t_o,\hat{t}_1] \to \mathbb{R}$ and $G : \mathbb{R}^n \times [t_o,\hat{t}_1] \to \mathbb{R}^k$ are continuous. Suppose f is continuously partially differentiable with respect to $x,\dot{x}$, and G with respect to x,t. Let $(\bar{x},\bar{t}_1) \in C \cap g^{-1}(0) \cap X_o$ be a local solution of (P). For brevity let $f(t) := f(\bar{x}(t),\dot{\bar{x}}(t),t)$. Define $\nabla_x f(t)$, $\nabla_{\dot{x}} f(t)$, $G(t)$, $G_x(t)$, $G_t(t)$ correspondingly. Let

$$\operatorname{rank}\,(G_x(\bar{t}_1)) = k.$$

Then one has

i) $$\nabla_{\dot{x}} f(t) = \int_{t_o}^{t} \nabla_x f(s)ds + \nabla_{\dot{x}} f(t_o) \text{ for all } t \in [t_o,\bar{t}_1].$$

ii) There exists a $\bar{y} \in \mathbb{R}^k$ with

$$\alpha)\ \nabla_{\dot{x}} f(\bar{t}_1) = -\, G_x(\bar{t}_1)^T\bar{y}$$

$$\beta)\ f(\bar{t}_1) - \nabla_{\dot{x}} f(\bar{t}_1)^T\dot{\bar{x}}(\bar{t}_1) + \bar{y}^T G_t(\bar{t}_1) = 0$$

Proof: I,g are continuously F-differentiable at $(\bar{x},\bar{t}_1)$, as one can show with a certain amount of effort, and one has

$$I'(\bar{x},\bar{t}_1)(h,s_1) = \int_{t_o}^{\bar{t}_1} \{\nabla_x f(t)^T h(t)+\nabla_{\dot{x}} f(t)^T\dot{h}(t)\}dt + f(\bar{t}_1)s_1$$

$$g'(\bar{x},\bar{t}_1)(h,s_1) = G_x(\bar{t}_1)h(\bar{t}_1) + (G_x(\bar{t}_1)\dot{\bar{x}}(\bar{t}_1) + G_t(\bar{t}_1))s_1$$

Furthermore $C - (\bar{x},\bar{t}_1) = \{(h,s) \in X : h(t_o) = 0\} = C(\bar{x},\bar{t}_1)$. Since $\operatorname{rank}\,(G_x(\bar{t}_1)) = k$ the constraint qualification

$$g'(\bar{x},\bar{t}_1)C(\bar{x},\bar{t}_1) = \mathbb{R}^k$$

is satisfied.

From the functional analytic "model theorem" given above we immediately conclude that there is a $\bar{y} \in \mathbb{R}^k$ with

$$(*)\quad \begin{aligned} &\int_{t_o}^{\bar{t}_1} \{\nabla_x f(t)^T h(t)+\nabla_{\dot{x}} f(t)^T\dot{h}(t)\}dt + f(\bar{t}_1)s_1 \\ &+ \bar{y}^T G_x(\bar{t}_1)h(t_1) + \bar{y}^T(G_x(\bar{t}_1)\dot{\bar{x}}(\bar{t}_1) + G_t(\bar{t}_1))s_1 = 0 \end{aligned}$$

for all $h \in C^1_{pc,n}[t_o,\bar{t}_1]$ with $h(t_o) = 0$ and all $s_1 \in \mathbb{R}$. Setting $s_1 = 0$ it then follows as we showed in Theorem 5.4.1 that the EULER equation in integrated form

$$\text{i)} \qquad \nabla_{\dot{x}} f(t) = \int_{t_o}^{t} \nabla_x f(s)ds + \nabla_{\dot{x}} f(t_o) \text{ for all } t \in [t_o, \bar{t}_1]$$

holds.

By partial integration in (*) and by using i) one gets

$$(\nabla_{\dot{x}} f(\bar{t}_1) + G_x(\bar{t}_1)^T \bar{y})^T h(\bar{t}_1) +$$
$$(f(\bar{t}_1) + \bar{y}^T (G_x(\bar{t}_1)\dot{\bar{x}}(\bar{t}_1) + G_t(\bar{t}_1)))s_1 = 0$$

for all $h \in C^1_{pc,n}[t_o, \bar{t}_1]$ with $h(t_o) = 0$ and all $s_1 \in \mathbb{R}$. The claim follows immediately.

Examples: 1) Let $G(x,t) = x - \varphi(t)$ with $\varphi \in C^1_n(-\infty,\infty)$. The transversality condition ii) in Theorem 5.4.2 now takes the form

$$f(\bar{t}_1) - \nabla_{\dot{x}} f(\bar{t}_1)^T (\dot{\bar{x}}(\bar{t}_1) - \dot{\varphi}(\bar{t}_1)) = 0$$

If moreover $n = 1$ and $f(x,\dot{x},t) = (1+\dot{x}^2)^{1/2}$, so that in the variable endpoint problem one is trying to find the shortest path from a point to a given curve, then the transversality condition looks as follows

$$(1+\dot{\bar{x}}^2(\bar{t}_1))^{1/2} - \frac{\dot{\bar{x}}(\bar{t}_1)}{(1+\dot{\bar{x}}^2(\bar{t}_1))^{1/2}} (\dot{\bar{x}}(\bar{t}_1) - \dot{\varphi}(\bar{t}_1)) = 0$$

$$\text{resp. } \dot{\bar{x}}(\bar{t}_1) = - \frac{1}{\dot{\varphi}(\bar{t}_1)}$$

and this is precisely the condition that $\bar{x}$ shall intersect the curve in a right angle.

2) Let $t_1 > t_o$ be fixed and $G(x,t) = t - t_1$. Then the problem described in Theorem 5.4.2 requires minimizing

$$I(x) := \int_{t_o}^{t_1} f(x(t),\dot{x}(t),t)dt$$

subject to $x(t_o) = x_o$, where the endpoint $x(t_1)$ is free. As a necessary optimality condition one gets in addition to the EULER equation the condition $\nabla_{\dot{x}} f(t_1) = 0$ (proof?). If one applies this to the problem of the brachystochrone with free endpoint, i.e. if $n = 1$ and

$$f(x,\dot{x},t) = \left(\frac{1+\dot{x}^2}{x_o-x}\right)^{1/2},$$

then one gets $\dot{\bar{x}}(t_1) = 0$. The desired curve is thus a cycloid which intersects the line $t - t_1 = 0$ perpendicularly. From the parameter representation

$$\left.\begin{aligned} t &= t_o + \frac{A}{2}(\varphi-\sin\varphi) \\ \bar{x} &= x_o - \frac{A}{2}(1-\cos\varphi) \end{aligned}\right\} \quad 0 \le \varphi \le \varphi_1$$

and $\dot{\bar{x}}(t_1) = 0$ we get that $\varphi_1 = \pi$ and $A = \dfrac{2(t_1-t_o)}{\pi}$.

3. Isoperimetric problems

By an isoperimetric variation problem one means the problem of minimizing

$$I(x) := \int_{t_o}^{t_1} f(x(t),\dot{x}(t),t)dt$$

subject to

$$g_i(x) := \int_{t_o}^{t_1} f_i(x(t),\dot{x}(t),t)dt - b_i = 0 \quad (i=1,\ldots,k)$$

and certain initial and endpoint conditions, say

$$x(t_o) = x_o, \; x(t_1) = x_1.$$

With the "model theorem" of the previous section one immediately has

5.4.3 Theorem: Let $X = C^1_{pc,n}[t_o,t_1]$. Suppose given the problem

$$(P)\quad \begin{aligned} &\text{Minimize } I(x) := \int_{t_o}^{t_1} f(x(t),\dot{x}(t),t)dt \text{ on } C \cap g^{-1}(0) \\ &\text{with } C := \{x \in X : x(t_o) = x_o,\ x(t_1) = x_1\} \\ &\text{and } g : X \to \mathbb{R}^k,\ g(x) = (g_1(x),\dots,g_k(x))^T \\ &\text{defined by } g_i(x) := \int_{t_o}^{t_1} f_i(x(t),\dot{x}(t),t)dt - b_i \quad (i=1,\dots,k). \end{aligned}$$

Here we suppose $x_o, x_1 \in \mathbb{R}^n$, $f, f_i : \mathbb{R}^n \times \mathbb{R}^n \times [t_o,t_1] \to \mathbb{R}$ $(i=1,\dots,k)$ are continuous and continuously partially differentiable with respect to $x, \dot{x}$.

Let $\bar{x} \in C \cap g^{-1}(0)$ be a local solution of (P). Again for brevity set $\nabla_x f(t) = \nabla_x f(\bar{x}(t),\dot{\bar{x}}(t),t)$ etc.

Then we have: there exist $\bar{y}_o \geq 0$, $\bar{y} \in \mathbb{R}^k$ with $(\bar{y}_o,\bar{y}) \neq (0,0)$ and

$$\bar{y}_o \nabla_{\dot{x}} f(t) + \sum_{i=1}^{k} \bar{y}_i \nabla_{\dot{x}} f_i(t) = \int_{t_o}^{t} (\bar{y}_o \nabla_x f(s) + \sum_{i=1}^{k} \bar{y}_i \nabla_x f_i(s))ds + \bar{y}_o \nabla_{\dot{x}} f(t_o) + \sum_{i=1}^{k} \bar{y}_i \nabla_{\dot{x}} f_i(t_o)$$

for all $t \in [t_o,t_1]$.

Example: Suppose given DIDO's problem (c.f. Example 3) in 1.2):

$$\text{Minimize } I(x) = \int_{-a}^{a} x(t)dt \text{ subject to}$$

$$x(-a) = x(a) = 0,\ \int_{-a}^{a} (1+\dot{x}^2(t))^{1/2}dt = 2l.$$

Suppose $0 < a < l < \pi a/2$. By the above theorem there exists $(\bar{y}_o,\bar{y}_1) \neq (0,0)$ with

$$\bar{y}_1 \frac{\dot{\bar{x}}(t)}{(1+\dot{\bar{x}}^2(t))^{1/2}} = \int_{-a}^{t} \bar{y}_o ds + \bar{y}_1 \frac{\dot{\bar{x}}(-a)}{(1+\dot{\bar{x}}^2(-a))^{1/2}}$$

for all $t \in [-a,a]$

We must have $\bar{y}_o \neq 0$ and thus without loss of generality $\bar{y}_o = 1$. For an appropriate constant c then

$$\dot{\bar{y}}_1^2 \frac{\dot{\bar{x}}^2(t)}{1+\dot{\bar{x}}^2(t)} = (t+a+c)^2 \text{ for all } t \in [-a,a]$$

and thus

$$\dot{\bar{x}}^2(t) = \frac{(t+a+c)^2}{\bar{y}_1^2-(t+a+c)^2}$$

$$= \left(\frac{d}{dt} (\bar{y}_1^2-(t+a+c)^2)^{1/2}\right)^2$$

Thus finally it follows

$$(\bar{x}(t)-d)^2 + (t+a+c)^2 = \bar{y}_1^2 \text{ for all } t \in [-a,a]$$

for a constant d, i.e. $\bar{x}$ is necessarily an arc of a circle. For computing $c,d,\bar{y}_1$ one has the side conditions $\bar{x}(-a) = 0 = \bar{x}(a)$,

$$\int_{-a}^{a} (1+\dot{\bar{x}}^2(t))^{1/2}dt = 2l.$$

Obviously one immediately gets $c = -a$; the determination of d and $\bar{y}_1$ is left as an exercise.

Remark: For the proof of Theorem 5.4.3 local smoothness assumptions on f, f_i naturally suffice (c.f. e.g. Theorem 5.4.1).

4. Optimal control theory

We consider a control problem given by the following data: a process, whose state is given as a function of time t by the trajectory $x(\cdot)$ and which can be controlled by the control function $u(\cdot)$ shall be described by the system of ordinary differential equations

$$\dot{x} = f(x,u,t).$$

The state of the process at an initial time t_o is given and fixed: $x(t_o) = x_o$. Starting from this initial state the process is to be controlled by $u(\cdot)$ such that its state $x(t_1)$ at a given terminal time $t_1 > t_o$ satisfies a terminal condition $G(x(t_1)) = 0$, the control condition $u(t) \in \Omega$ is satisfied for all $t \in [t_o, t_1]$ and an objective function of the form

$$I(x,u) = F(x(t_1)) + \int_{t_o}^{t_1} f^o(x(t),u(t),t)\,dt$$

is minimal.

This time we skip the formulation of a functional analytic model theorem but remind the reader of the definition of the Banach space $W_n^{1,\infty}[t_o,t_1]$ (c.f. 5.1 and 5.2):

$$W_n^{1,\infty}[t_o,t_1] = \{x \in C_n[t_o,t_1] : x(t) = x(t_o) + \int_{t_o}^{t} \dot{x}(s)\,ds \quad \text{with } \dot{x} \in L_n^{\infty}[t_o,t_1]\}$$

with $\| x \| = \max\,(\| x \|_\infty, \| \dot{x} \|_\infty)$ for $x \in W_n^{1,\infty}[t_o,t_1]$.

5.4.4 Theorem (Local PONTRYAGIN maximum principle): Let $X := W_n^{1,\infty}[t_o,t_1]$, $U := L_m^{\infty}[t_o,t_1]$. Suppose given the problem

$$\text{Minimize } I(x,u) := F(x(t_1)) + \int_{t_o}^{t_1} f^o(x(t),u(t),t)\,dt$$

on $C \cap g^{-1}(0,0)$ with

(P) $C := \{(x,u) \in X \times U : x(t_o) = x_o, u(t) \in \Omega \text{ a.e. on } [t_o,t_1]\}$

$g : X \times U \to L_n^{\infty}[t_o,t_1] \times \mathbb{R}^k$ defined by

$$g(x,u) = (\dot{x}(\cdot) - f(x(\cdot),u(\cdot),\cdot), G(x(t_1))).$$

$$f : \mathbb{R}^n \times \mathbb{R}^m \times [t_o,t_1] \to \mathbb{R}^n \text{ and } f^o : \mathbb{R}^n \times \mathbb{R}^m \times [t_o,t_1] \to \mathbb{R}$$

are to be continuous and continuously partially differentiable with respect to x,u, $F : \mathbb{R}^n \to \mathbb{R}$ and $G : \mathbb{R}^n \to \mathbb{R}^k$ continuously partially differentiable. $\Omega \subset \mathbb{R}^m$ shall be nonempty, convex and closed.

Let $(\bar{x},\bar{u}) \in C \cap g^{-1}(0,0)$ be a local solution of (P) and suppose rank $G_x(\bar{x}(t_1)) = k$. Again for brevity let

$$\nabla_x f^o(t) := \nabla_x f^o(\bar{x}(t),\bar{u}(t),t), \nabla_u f^o(t) := \nabla_u f^o(\bar{x}(t),\bar{u}(t),t),$$

$\nabla F(t_1) := \nabla F(\bar{x}(t_1))$. Analogously we let $f_x(t)$, $f_u(t)$ denote the Jacobi matrices of f with respect to x resp. u at $(\bar{x}(t),\bar{u}(t),t)$ and $G_x(t_1)$ the Jacobi matrix of G at $\bar{x}(t_1)$.

Then there exist $\lambda_o \geq 0$, $\eta \in W_n^{1,\infty}[t_o,t_1]$ and $\mu \in \mathbb{R}^k$ with $(\lambda_o,\eta) \neq (0,0)$ and

1. $$-\dot{\eta}(t) = f_x^T(t)\eta(t) - \lambda_o \nabla_x f^o(t) \text{ a.e. on } [t_o,t_1]$$

(adjoint equation)

2. $$-\eta(t_1) = G_x^T(t_1)\mu + \lambda_o \nabla F(t_1)$$

(transversality condition)

3. For a.a. $t \in [t_o,t_1]$ one has

$$(u-\bar{u}(t))^T(f_u(t)^T\eta(t) - \lambda_o \nabla_u f^o(t)) \leq 0 \text{ for all } u \in \Omega.$$

(local PONTRYAGIN maximum principle).

Proof: The idea of the proof consists in first showing that the hypotheses (V) i) - v) of Theorem 5.3.2 with $Y := L_n^\infty[t_o,t_1] \times \mathbb{R}^k$ and $K = \{(0,0)\}$ hold. Since the constraint qualification $g'(\bar{x},\bar{u})C(\bar{x},\bar{u}) = Y$ will not necessarily be satisfied without further hypotheses and since it is difficult to prove the closure of $g'(\bar{x},\bar{u})C(\bar{x},\bar{u})$ (and subsequently to apply Theorem 5.3.6), we shall show icr $g'(\bar{x},\bar{u})C(\bar{x},\bar{u}) \neq \emptyset$ and apply Theorem 5.3.7.

1) Proof of the hypotheses (V) i) - v) of Theorem 5.3.2.

(V) i) $X \times U = W_n^{1,\infty}[t_o,t_1] \times L_m^\infty[t_o,t_1]$ is a Banach space; as norm on this product space one chooses e.g.

$$\| (x,u) \| = \max(\| x \|, \| u \|_\infty).$$

ii) $C := \{(x,u) \in X \times U : x(t_o) = x_o, u(t) \in \Omega$ a.e. on $[t_o,t_1]\}$ is nonempty, convex and closed. For: Since Ω is nonempty and convex, C is evidently nonempty and convex. The proof of the closure of C is somewhat more difficult:

If $\{u_k\} \subset L_m^\infty[t_o,t_1]$ converges to some $u \in L_m^\infty[t_o,t_1]$ i.e.

$$\varepsilon_k := \| u-u \|_\infty \to 0,$$

then there exists a set $E_k \subset [t_o,t_1]$ of measure 0 with

$$|u_k(t)-u(t)| \leq \varepsilon_k \text{ for all } t \in [t_o,t_1] \setminus E_k.$$

If one sets $E := \bigcup_{k=1}^{\infty} E_k$, then E is also a set of measure 0 and

$$\lim_{k\to\infty} u_k(t) = u(t) \text{ for all } t \in [t_o,t_1] \setminus E.$$

But from the closure of Ω it then immediately follows that C is closed.

iii) $L_n^\infty[t_o,t_1] \times \mathbb{R}^k$ is a Banach space.

iv) $K := \{(0,0)\} \subset L_n^\infty[t_o,t_1] \times \mathbb{R}^k$ is trivially a nonempty, closed convex cone.

v) I is F-differentiable at $(\bar{x},\bar{u})$ with

$$I'(\bar{x},\bar{u})(h,v) = \nabla F(t_1)^T h(t_1) + \int_{t_o}^{t_1} \{\nabla_x f^o(t)^T h(t) + \nabla_u f^o(t)^T v(t)\}dt$$

and g is continuously F-differentiable at $(\bar{x},\bar{u})$ with

$$g'(\bar{x},\bar{u})(h,v) = (\dot{h}-f_x(\cdot)h-f_u(\cdot)v, G_x(t_1)h(t_1)).$$

We shall skip the proof of this fact. It can be found in

KIRSCH-WARTH-WERNER [42, p. 94]; see also Example 2) in 5.1.

2) icr $g'(\bar{x},\bar{u})C(\bar{x},\bar{u}) \neq \emptyset$. Obviously

$$C(\bar{x},\bar{u}) = \{(h,v) \in X \times U : h(t_o) = 0, v = \lambda(u-\bar{u}) \text{ with } \lambda \geq 0, u(t) \in \Omega \text{ a.e. on } [t_o,t_1]\},$$

is the cone generated by $C - (\bar{x},\bar{u})$. Since $g'(\bar{x},\bar{u})$ is linear, it suffices to show that icr $C(\bar{x},\bar{u}) \neq \emptyset$. Since $\Omega \subset \mathbb{R}^m$ is non-empty and convex there is a $u_o \in \Omega$ and an $\varepsilon > 0$ with

$$B[u_o;\varepsilon] \cap \operatorname{aff}(\Omega) \subset \Omega \quad \text{(proof?)}.$$

We want to show that $(0,u_o-\bar{u}) \in \operatorname{icr} C(\bar{x},\bar{u})$.

α) $$\operatorname{aff} C(\bar{x},\bar{u}) \subset \{(h,\lambda(u-\bar{u})) \in X \times U : h(t_o) = 0, \lambda \geq 0 \text{ and } u(t) \in \operatorname{aff}(\Omega) \text{ a.e. on } [t_o,t_1]\},$$

for the set on the right hand side contains $C(\bar{x},\bar{u})$ and is an affine subspace of $X \times U$.

β) Suppose $(h,\lambda(u-\bar{u})) \in X \times U$ with $h(t_o) = 0$, $\lambda \geq 0$ and $u(t) \in \operatorname{aff}(\Omega)$ are given and fixed. Choose $\tau > 0$ so small that

$$1 - \tau + \lambda\tau > 0 \text{ and } \frac{\lambda\tau}{1-\tau+\lambda\tau} \| u-u_o \|_\infty \leq \varepsilon.$$

Then

$$(0,u_o-\bar{u}) + \tau((h,\lambda(u-\bar{u}))-(0,u_o-\bar{u})) =$$

$$\left(\tau h, (1-\tau+\lambda\tau)\left\{\frac{1-\tau}{1-\tau+\lambda\tau} u_o + \frac{\lambda\tau}{1-\tau+\lambda\tau} u-\bar{u}\right\}\right) \in C(\bar{x},\bar{u})$$

for

$$\frac{1-\tau}{1-\tau+\lambda\tau} u_o + \frac{\lambda\tau}{1-\tau+\lambda\tau} u(t) \in \operatorname{aff}(\Omega) \cap B[u_o,\varepsilon] \subset \Omega \quad \text{a.e. on } [t_o,t_1].$$

Thus $(0,u_o-\bar{u}) \in \operatorname{icr} C(\bar{x},\bar{u})$ resp.

$$g'(\bar{x},\bar{u})(0,u_o-\bar{u}) \in \operatorname{icr} g'(\bar{x},\bar{u})C(\bar{x},\bar{u})$$

and 2) is proved.

3) Theorem 5.3.7 gives the existence of

$$(\lambda_o,l,\mu) \in \mathbb{R} \times (L_n^\infty[t_o,t_1])' \times \mathbb{R}^k \setminus \{(0,0,0)\} \text{ with } \lambda_o \geq 0 \text{ and}$$

$$\lambda_o \nabla F(t_1)^T h(t_1) + \lambda_o \int_{t_o}^{t_1} \{\nabla_x f^o(t)^T h(t) + \nabla_u f^o(t)^T v(t)\} dt$$

$$+ \langle l, \dot{h} - f_x(\cdot)h - f_u(\cdot)v\rangle + \mu^T G_x(t_1)h(t_1) \geq 0$$

for all $(h,v) \in C(\bar{x},\bar{u})$.

In particular

$$(*) \quad (\lambda_o \nabla F(t_1) + G_x^T(t_1)\mu)^T h(t_1) + \lambda_o \int_{t_o}^{t_1} \nabla_x f^o(t)^T h(t) dt$$

$$+\langle l, \dot{h} - f_x(\cdot)h \rangle = 0 \text{ for all } h \in W_n^{1,\infty}[t_o,t_1] \text{ with } h(t_o) = 0$$

and

$$(**) \quad \lambda_o \int_{t_o}^{t_1} \nabla_u f^o(t)^T (u(t)-\bar{u}(t)) dt - \langle l, f_u(\cdot)(u-\bar{u})\rangle \geq 0$$

for all $u \in L_m^\infty[t_o,t_1]$ with $u(t) \in \Omega$ a.e. on $[t_o,t_1]$.

4) "Calculation" of $l \in (L_n^\infty[t_o,t_1])'$.
Let $y \in L_n^\infty[t_o,t_1]$ be arbitrarily given. The initial value problem

$$\dot{h} - f_x(t)h = y(t) \;,\; h(t_o) = 0$$

has a unique solution $h \in W_n^{1,\infty}[t_o,t_1]$ which can be given in the form

$$h(t) = \Phi(t) \int_{t_o}^{t} \Phi^{-1}(s) y(s) ds.$$

Here Φ is the fundamental system for $\dot{h} - f_x(t)h$ normalized by

$\Phi(t_o) = I$. Then from (*) one gets

$$<l,y> = -(\lambda_o \nabla F(t_1)+G_x^T(t_1)\mu)^T \Phi(t_1) \int_{t_o}^{t_1} \Phi^{-1}(t)y(t)dt$$

$$- \lambda_o \int_{t_o}^{t_1} \nabla_x f^o(t)^T \Phi(t) \int_{t_o}^{t} \Phi^{-1}(s)y(s)ds\, dt$$

$$= -(\lambda_o \nabla F(t_1)+G_x^T(t_1)\mu)^T \Phi(t_1) \int_{t_o}^{t_1} \Phi^{-1}(t)y(t)dt$$

$$+ \lambda_o \int_{t_o}^{t_1} \int_{t_1}^{t} \nabla_x f^o(s)^T \Phi(s)ds \Phi^{-1}(t)y(t)dt$$

$$= \int_{t_o}^{t_1} \eta(t)^T y(t)dt,$$

where $\eta(t) = -\Phi^T(t)^{-1}\Phi^T(t_1)(\lambda_o \nabla F(t_1)+G_x^T(t_1)\mu)$

$$+ \Phi^T(t)^{-1} \int_{t_1}^{t} \Phi^T(s)\lambda_o \nabla_x f^o(s)ds$$

is the solution of the adjoint equation

$$-\dot{\eta} = f_x(t)^T \eta - \lambda_o \nabla_x f^o(t)$$

with the end condition

$$-\eta(t_1) = G_x^T(t_1)\mu + \lambda_o \nabla F(t_1).$$

Furthermore $(\lambda_o,\eta) \in \mathbb{R} \times W_n^{1,\infty}[t_o,t_1] \setminus \{(0,0)\}$, for if $(\lambda_o,\eta) = (0,0)$, then one would have $(\lambda_o,l) = (0,0)$ and $0 = G_x^T(t_1)\mu$, and hence $\mu = 0$ since rank $G_x(t_1) = k$, a contradiction.

5) The local PONTRYAGIN maximum principle.
From (**) and

$$<l,y> = \int_{t_o}^{t_1} \eta(t)^T y(t)dt$$

we get

$$\int_{t_o}^{t_1} (\lambda_o \nabla_u f^o(t) - f_u(t)^T \eta(t))^T (u(t) - \bar{u}(t))dt \geq 0$$

resp.

$$\int_{t_o}^{t_1} (u(t) - \bar{u}(t))^T (f_u(t)^T \eta(t) - \lambda_o \nabla_u f^o(t))dt \leq 0$$

for all $u \in L_m^\infty[t_o,t_1]$ with $u(t) \in \Omega$ a.e. on $[t_o,t_1]$.

Now let $u \in \Omega$ be arbitrary and $t \in (t_o,t_1)$. For sufficiently small $\varepsilon > 0$ define $u_\varepsilon \in L_m^\infty[t_o,t_1]$ by

$$u_\varepsilon(s) = \begin{cases} u & \text{for } s \in [t-\varepsilon,t] \\ \bar{u}(s) & \text{for } s \in [t_o,t_1] \setminus [t-\varepsilon,t]. \end{cases}$$

Then

$$\frac{1}{\varepsilon} \int_{t_o}^{t_1} (u_\varepsilon(s) - \bar{u}(s))^T (f_u(s)^T \eta(s) - \lambda_o \nabla_u f^o(s))ds$$

$$= \frac{1}{\varepsilon} \int_{t-\varepsilon}^{t} (u - \bar{u}(s))^T (f_u(s)^T \eta(s) - \lambda_o \nabla_u f^o(s))ds \leq 0.$$

Since moreover for a.a. $t \in (t_o,t_1)$

$$\lim_{\varepsilon \to 0+} \frac{1}{\varepsilon} \int_{t-\varepsilon}^{t} (u - \bar{u}(s))^T (f_u(s)^T \eta(s) - \lambda_o \nabla_u f^o(s))ds =$$

$$(u - \bar{u}(t))^T (f_u(t)^T \eta(t) - \lambda_o \nabla_u f^o(t))$$

the claim follows, namely that

$$\max_{u \in \Omega} u^T (f_u(t)^T \eta(t) - \lambda_o \nabla_u f^o(t))$$

$$= \bar{u}(t)^T (f_u(t)^T \eta(t) - \lambda_o \nabla_u f^o(t)) \text{ a.e. on } [t_o,t_1].$$

<u>Remark</u>: If for the optimal control problem (P) in Theorem 5.4.4 one defines the so-called HAMILTON function

$$H : \mathbb{R}^n \times \mathbb{R}^m \times [t_o,t_1] \times \mathbb{R}^n \times \mathbb{R} \to \mathbb{R}$$

by

$$H(x,u,t,\eta,\lambda_o) := \eta^T f(x,u,t) - \lambda_o f^o(x,u,t).$$

then one can write the process equation $\dot{x} = f(x,u,t)$ in the form $\dot{x} = \nabla_\eta H$ and the adjoint equation reads $-\dot{\eta} = \nabla_x H$. Then one can formulate the statement of the local PONTRYAGIN maximum principle as follows:

$$(u-\bar{u}(t))^T \nabla_u H(\bar{x}(t),\bar{u}(t),t,\eta(t),\lambda_o) \leq 0$$

for all $u \in \Omega$ and a.a. $t \in [t_o,t_1]$.

If now H is concave with respect to the variable u (i.e. -H is convex, e.g. $f(x,u,t) = A(t)x + B(t)u$ linear with respect to u and f^o convex with respect to u), then

$$\begin{aligned} 0 &\leq (u-\bar{u}(t))^T(-\nabla_u H(\bar{x}(t),\bar{u}(t),t,\eta(t),\lambda_o)) \\ &\leq - H(\bar{x}(t),u,t,\eta(t),\lambda_o) + H(\bar{x}(t),\bar{u}(t),t,\eta(t),\lambda_o), \end{aligned}$$

and hence

$$\max_{u\in\Omega} H(\bar{x}(t),u,t,\eta(t),\lambda_o) = H(\bar{x}(t),\bar{u}(t),t,\eta(t),\lambda_o)$$

a.e. on $[t_o,t_1]$.

Surprisingly this holds even without the convexity of Ω and of - H with respect to u. This is then called the (global) PONTRYAGIN maximum principle. We cannot go into the proof. The reader is referred e.g. to IOFFE-TICHOMIRIV [37, p. 126], FLEMING-RISHEL [24, p. 27].

Entirely analogously to Theorem 5.4.4 one can get necessary optimality conditions for optimal control problems on a variable time interval (c.f. e.g. Theorem 5.4.2). We shall, however, not go into this topic and excuse ourselves by pointing out once more that the exposition in this section is only intended to illustrate the use of abstract necessary optimality conditions and in no sense to be an exhaustive treatment of the theory of necessary conditions in the calculus of variations and in optimal control theory.

5.5 Necessary and sufficient optimality conditions of second order

It is well known that if a twice continuously differentiable function $f : \mathbb{R}^n \to \mathbb{R}$ has a local minimum at a point $\bar{x}$ then

$$\nabla f(\bar{x}) = \left(\frac{\partial f}{\partial x_1}(\bar{x}), \ldots, \frac{\partial f}{\partial x_n}(\bar{x})\right)^T = 0$$

(necessary optimality condition of first order) and

$$\nabla^2 f(\bar{x}) = \left(\frac{\partial^2 f}{\partial x_i \partial x_j}(\bar{x})\right) \text{ is positive semidefinite}$$

(necessary optimality condition of second order).

If conversely

$$\nabla f(\bar{x}) = 0 \text{ and } \nabla^2 f(\bar{x}) \text{ is positive definite,}$$

then $\bar{x}$ is an isolated local minimum of f (sufficient optimality condition of second order). Clearly one might try to carry over these statements to constrained optimization problems and in the process try if possible to handle infinite dimensional problems. For this purpose one of course needs second derivatives. The conceptually simplest definition is

5.5.1 Definition: Let X,Y be normed linear spaces. A map $g : X \to Y$ is called twice FRECHET differentiable at $x \in X$ if

1. g is continuously F-differentiable at x with F-differential $g'(x)$.
2. There exists a map $B : X \times X \to Y$ which is bilinear ($B(\cdot,x_2)$ and $B(x_1,\cdot)$ are linear maps from X to Y for arbitrary $x_1,x_2 \in X$) and continuous and for which

$$\lim_{\|h\|\to 0} \frac{\| g(x+h)-g(x)-g'(x)h-\frac{1}{2}B(h,h) \|}{\|h\|^2} = 0.$$

$B = g''(x)$ is called the second FRECHET derivative of g at x.

Remark: Exactly as in the remark following Definition 5.1.4, the definition of the F-differential, one can see that the second F-derivative $g''(x)$ is uniquely determined.

Examples: 1) Suppose $f : \mathbb{R}^n \to \mathbb{R}$ is twice continuously partially differentiable at x. Then f is twice F-differentiable at x and

$$f''(x)(h,h) = h^T \nabla^2 f(x) h,$$

where

$$\nabla^2 f(x) = \left(\frac{\partial^2 f}{\partial x_i \partial x_j} (x) \right)$$

is the Hessian of f at x.

2) Let $I : C^1[t_o,t_1] \to \mathbb{R}$ be defined by

$$I(x) := \int_{t_o}^{t_1} f(x(t),\dot{x}(t),t)dt.$$

Let $f : \mathbb{R} \times \mathbb{R} \times [t_o,t_1] \to \mathbb{R}$ be continuous and twice continuously partially differentiable in $x,\dot{x}$. Then I is twice F-differentiable at $x \in C^1[t_o,t_1]$ and

$$I''(x)(h,h) =$$

$$\int_{t_o}^{t_1} \{f_{xx}(t)h(t)^2+2f_{x\dot{x}}(t)h(t)\dot{h}(t)+f_{\dot{x}\dot{x}}(t)\dot{h}(t)^2\}dt.$$

Here we have abbreviated

$$f_{xx}(t) = \frac{\partial^2 f}{\partial x^2} (x(t),\dot{x}(t),t), f_{x\dot{x}}(t), f_{\dot{x}\dot{x}}(t) \text{ analogously.}$$

For the sake of simplicity in what follows we only consider optimization problems without explicit constraints and prove

5.5.2 Theorem: Suppose given the optimization problem

(P) Minimize $f(x)$ on $M := \{x \in X : g(x) \in - K\}$.

We assume

(V) i) $f : X \to \mathbb{R}$, X a Banach space.

ii) $g : X \to Y$, Y a Banach space.

iii) $K \subset Y$ is a nonempty, closed convex cone.

iv) $\bar{x} \in M$ is a local solution of (P), f and g are twice F-differentiable at $\bar{x}$.

Furthermore we suppose the constraint qualification

(CQ) $\quad g'(\bar{x})X + K(-g(\bar{x})) = Y$

holds. Let $\bar{y}^* \in Y^*$ be a Lagrange multiplier for (P) at $\bar{x}$, which exists by Theorem 5.3.2, (i.e. $\bar{y}^* \in K^+$, $\langle \bar{y}^*, g(\bar{x})\rangle = 0$ and

$$f'(\bar{x}) + \bar{y}^* \circ g'(\bar{x}) = 0).$$

If one defines

$$\bar{K} := \{z \in K : \langle \bar{y}^*, z\rangle = 0\} \text{ and}$$

$$\bar{M} := \{x \in X : g(x) \in -\bar{K}\}, \text{ then}$$

$$(f''(\bar{x}) + \bar{y}^* \circ g''(\bar{x}))(h,h) \geq 0 \text{ for all } h \in T(\bar{M};\bar{x}).$$

If

$(\overline{CQ})$ $\quad g'(\bar{x})X + \bar{K}(-g(\bar{x})) = Y$

then

$$(f''(\bar{x}) + \bar{y}^* \circ g''(\bar{x}))(h,h) \geq 0 \text{ for all } h \in L(\bar{M};\bar{x}).$$

<u>Proof</u>: Suppose $h \in T(\bar{M};\bar{x})$. By definition of the tangent cone (Definition 5.2.1) there exist sequences $\{t_j\} \subset \mathbb{R}_+$, $\{r_j\} \subset X$ with

i) $\quad \bar{x} + t_j h + r_j \in \bar{M}$, i.e. $g(\bar{x} + t_j h + r_j) \in -\bar{K}$.

ii) $\quad t_j \to 0$, $r_j/t_j \to 0$.

If one sets $F(x) := f(x) + \bar{y}^* \circ g(x)$, then

$$F(\bar{x}+t_j h+r_j) - F(\bar{x}) = f(\bar{x}+t_j h+r_j) - f(\bar{x}) \geq 0$$

for all sufficiently large j.

Since $F'(\bar{x}) = f'(\bar{x}) + \bar{y}^* \circ g'(\bar{x}) = 0$ we thus have

$$0 \leq \lim_{j\to\infty} \frac{F(\bar{x}+t_j h+r_j)-F(\bar{x})}{t_j^2} = \frac{1}{2} F''(\bar{x})(h,h)$$

for all $h \in T(\bar{M};\bar{x})$.

If the constraint qualification

$$(\overline{CQ}) \qquad g'(\bar{x})X + \bar{K}(-g(\bar{x})) = Y$$

holds, then by the Theorem of LYUSTERNIK (Theorem 5.2.5) one has

$$L(\bar{M};\bar{x}) = \{h \in X : g'(\bar{x})h \in -\bar{K}(-g(\bar{x}))\} \subset T(\bar{M};\bar{x})$$

and the theorem is proved.

A specialization to finite dimensional optimization problems gives

<u>5.5.3 Theorem</u>: Suppose given the optimization problem

(P) Minimize $f(x)$ on

$$M := \{x \in \mathbb{R}^n : g_i(x) \leq 0 \ (i=1,\ldots,m),$$
$$g_i(x) = 0 \ (i=m+1,\ldots,m+k)\}.$$

Let $f, g_i : \mathbb{R}^n \to \mathbb{R}$ $(i=1,\ldots,m+k)$ be twice continuously differentiable at a local solution $\bar{x} \in M$ of (P). Let

$$I(\bar{x}) := \{i \in \{1,\ldots,m\} : g_i(\bar{x}) = 0\}.$$

Further suppose that the vectors $\nabla g_i(\bar{x})$ $(i \in I(\bar{x}))$,

$\nabla g_{m+1}(\bar{x}),\dots,\nabla g_{m+k}(\bar{x})$ are linearly independent. Then there exists a $\bar{y} \in \mathbb{R}^{m+k}$ with

i) $\quad \bar{y}_i \geq 0 \quad (i,\dots,m)$

ii) $\quad \bar{y}_i g_i(\bar{x}) = 0 \quad (i=1,\dots,m)$

iii) $\quad \nabla f(\bar{x}) + \sum_{i=1}^{m+k} \bar{y}_i \nabla g_i(\bar{x}) = 0$ and

iv) $\quad h^T\left(\nabla^2 f(\bar{x}) + \sum_{i=1}^{m+k} \bar{y}_i \nabla^2 g_i(\bar{x})\right)h \geq 0$ for all

$$h \in L := \{h \in \mathbb{R}^n : \nabla g_i(\bar{x})^T h = 0 \text{ for } i \in I(\bar{x}) \text{ and } i = m+1,\dots,m+k\}.$$

Proof: We want to apply Theorem 5.5.2 and thus set $X = \mathbb{R}^n$, $Y = \mathbb{R}^{m+k}$ and

$$K := \{y = (y_i) \in \mathbb{R}^{m+k} : y_i \geq 0 \ (i=1,\dots,m),\ y_i = 0 \ (i=m+1,\dots,m+k)\}.$$

Let $g(x) := (g_i(x))$. The constraint qualification

(CQ) $\quad g'(\bar{x})X + K(-g(\bar{x})) = Y$

is satisfied if

a) $\exists\, h \in \mathbb{R}^n$ with $\nabla g_i(\bar{x})^T h < 0 \ (i \in I(\bar{x}))$ and

$\nabla g_i(\bar{x})^T h = 0$ for $i = m+1,\dots,m+k$;

b) $\nabla g_{m+1}(\bar{x}),\dots,\nabla g_{m+k}(\bar{x})$ are linearly independent

(c.f. Example following Lemma 5.3.4). Since $\nabla g_i(\bar{x})$ $(i \in I(\bar{x}))$, $\nabla g_{m+1}(\bar{x}),\dots,\nabla g_{m+k}(\bar{x})$ were linearly independent by hypothesis, there exists an $h \in \mathbb{R}^n$ with $\nabla g_i(\bar{x})^T h = -1$ $(i \in I(\bar{x}))$ and $\nabla g_i(\bar{x})^T h = 0$ $(i=m+1,\dots,m+k)$ - i.e. (CQ) is satisfied. Let $\bar{y} \in \mathbb{R}^{m+k}$ be a Lagrange multiplier for (P) at $\bar{x}$. $\bar{y}$ satisfies i), ii), iii). As in Theorem 5.5.2 we define

$$\overline{K} := \{y \in K : \overline{y}^T y = 0\}$$

$$= \{y = (y_i) \in \mathbb{R}^{m+k} : y_i \geq 0 \ (i=1,\dots,m),\ y_i = 0$$

$$\text{for all } i \in \{1,\dots,m\} \text{ with } \overline{y}_i > 0 \text{ and } i = m+1,\dots,m+k\}.$$

The constraint qualification

$$(\overline{CQ}) \quad g'(\overline{x})X + \overline{K}(-g(\overline{x})) = Y$$

is also satisfied. To see this one must show:

if $y \in \mathbb{R}^{m+k}$ is given, then there exist $h \in \mathbb{R}^n$, $\lambda \geq 0$ and $z \in \overline{K}$ with

$$\nabla g_i(\overline{x})^T h + \lambda(z_i + g_i(\overline{x})) = y \quad (i=1,\dots,m+k).$$

To this end first choose an h as solution of $\nabla g_i(\overline{x})^T h = y_i$ ($i \in I(\overline{x})$ and $i = m+1,\dots,m+k$). Then choose a $\lambda > 0$ so large that

$$z_i := -g_i(\overline{x}) + \frac{1}{\lambda}(y_i - \nabla g_i(\overline{x})^T h) \geq 0 \text{ for } i \in \{1,\dots,m\} \setminus I(\overline{x}).$$

If one sets $z_i = 0$ for $i \in I(\overline{x})$ and $i = m+1,\dots,m+k$, then $z = (z_i)$ is in $\overline{K}$ and the constraint qualification $(\overline{CQ})$ is satisfied. If one then notes that

$$L := \{h \in \mathbb{R}^n : \nabla g_i(\overline{x})^T h = 0 \text{ for } i \in I(\overline{x}) \text{ and } i = m+1,\dots,m+k\}$$

$$\subset L(\overline{M};\overline{x}) = \{h \in \mathbb{R}^n : g'(\overline{x})h \in -\overline{K}(-g(\overline{x}))\},$$

then the claim of the theorem follows from Theorem 5.5.2.

Remark: If the g_i are affine linear as in Theorem 5.3.5, i.e.

$$g_i(x) = a^{iT}x - \beta_i \quad (i=1,\dots,m+k),$$

then one can do without the constraint qualification in Theorem 5.5.3 (proof?).

We conclude this section by going into the topic of sufficient conditions of second order but restrict ourselves to finite dimensional problems; the general case is treated e.g. in MAURER-ZOWE [59].

5.5.4 Theorem: Suppose given the optimization problem

(P) Minimize $f(x)$ on

$$M := \{x \in \mathbb{R}^n : g_i(x) \le 0 \ (i=1,\dots,m),$$
$$g_i(x) = 0 \ (i=m+1,\dots,m+k)\}.$$

Suppose f, g_i $(i=1,\dots,m+k)$ are twice continuously partially differentiable at $\bar{x} \in M$. Suppose there exists a $\bar{y} \in \mathbb{R}^{m+k}$ with

i) $\bar{y}_i \ge 0 \ (i=1,\dots,m)$

ii) $\bar{y}_i g_i(\bar{x}) = 0 \ (i=1,\dots,m)$

iii) $\nabla f(\bar{x}) + \sum_{i=1}^{m+k} \bar{y}_i \nabla g_i(\bar{x}) = 0$

iv) Let $L' := \{h \in \mathbb{R}^n : \nabla g_i(\bar{x})^T h = 0$ for $i \in I'(\bar{x})$,

$$\nabla g_i(\bar{x})^T h = 0 \ (i=m+1,\dots,m+k)\}$$

with $I'(\bar{x}) := \{i \in \{1,\dots,m\} : \bar{y}_i > 0\}$.

Then we require

$$h^T(\nabla^2 f(\bar{x}) + \sum_{i=1}^{m+k} \bar{y}_i \nabla^2 g_i(x))h > 0 \text{ for } h \in L' \setminus \{0\}.$$

Then $\bar{x}$ is a strict local solution of (P), i.e. there is a sphere $B[\bar{x};\varepsilon]$ around $\bar{x}$ with radius $\varepsilon > 0$ such that $f(\bar{x}) < f(x)$ for all $x \in B[\bar{x};\varepsilon] \cap M$ with $x \neq \bar{x}$.

Proof: Suppose $\bar{x}$ were not a strict local solution of (P). Then there exists a sequence $\{x^j\} \subset M \setminus \{\bar{x}\}$ with $x^j \to \bar{x}$ and $f(x^j) \le f(\bar{x})$. Let us represent x^j in the form

$$x^j = \bar{x} + t_j h^j \text{ with } t_j > 0 \text{ and } |h^j| = 1.$$

Since $x^j \to \bar{x}$ we have $t_j \to 0$. Let h be an accumulation point of $\{h^j\}$. Then $|h| = 1$ and without loss of generality we may assume $h^j \to h$.

a) $\nabla f(\bar{x})^T h \le 0$, $\nabla g_i(\bar{x})^T h = 0$ for $i = m+1,\dots,m+k$ and $\nabla g_i(\bar{x})^T h \le 0$ for $i \in I(\bar{x}) = \{i \in \{1,\dots,m\} : g_i(\bar{x}) = 0\}$. For:

$$0 \ge \frac{f(x^j)-f(\bar{x})}{t_j} = \frac{f(\bar{x}+t_j h^j)-f(\bar{x})}{t_j} \to \nabla f(\bar{x})^T h.$$

For $i = m+1,\dots,m+k$ we have

$$0 = \frac{g_i(x^j)-g_i(\bar{x})}{t_j} = \frac{g_i(\bar{x}+t_j h^j)-g_i(\bar{x})}{t_j} \to \nabla g_i(\bar{x})^T h .$$

For $i \in I(\bar{x})$ we have

$$0 \ge \frac{g_i(x^j)-g_i(\bar{x})}{t_j} = \frac{g_i(\bar{x}+t_j h^j)-g_i(\bar{x})}{t_j} \to \nabla g_i(\bar{x})^T h.$$

b) We distinguish two cases:

1. Suppose there exists an $i \in I'(\bar{x}) = \{i \in I(\bar{x}) : \bar{y}_i > 0\}$ with $\nabla g_i(\bar{x})^T h < 0$. Then by iii) we have

$$0 \ge \nabla f(\bar{x})^T h = - \sum_{i=1}^{m+k} \bar{y}_i \nabla g_i(\bar{x})^T h$$

$$= - \sum_{i \in I'(\bar{x})} \bar{y}_i \nabla g_i(\bar{x})^T h > 0.$$

   a contradiction.

2. Suppose for all $i \in I'(\bar{x})$ one has $\nabla g_i(\bar{x})^T h = 0$. Then $h \in L'$. Now let $i \in I'(\bar{x}) \cup \{m+1,\dots,m+k\}$. Then

$$0 = g_i(x^j) = g_i(\bar{x}) + t_j \nabla g_i(\bar{x})^T h^j + \frac{1}{2} t_j^2 h^{jT} \nabla^2 g_i(\xi_i^j) h^j$$

   and

$$0 \ge f(x^j) - f(\bar{x}) = t_j \nabla f(\bar{x})^T h^j + \frac{1}{2} t_j^2 h^{jT} \nabla^2 f(\eta^j) h^j$$

with $\xi_i^j, \eta^j \in (x^j, \bar{x})$. Thus

$$0 \geq t_j \left\{ \nabla f(\bar{x}) + \sum_{i=1}^{m+k} \bar{y}_i \nabla g_i(\bar{x}) \right\}^T h^j$$
$$+ \frac{1}{2} t_j^2 h^{jT} \left\{ \nabla^2 f(\eta^j) + \sum_{i=1}^{m+k} \bar{y}_i \nabla^2 g_i(\xi_i^j) \right\} h^j.$$

The first summand vanishes by iii). Hence

$$h^{jT} \left\{ \nabla^2 f(\eta^j) + \sum_{i=1}^{m+k} \bar{y}_i \nabla^2 g_i(\xi_i^j) \right\} h^j \leq 0.$$

Letting j tend to ∞ and remembering that $h^j \to h$ and $\eta^j, \xi_i^j \to \bar{x}$ we get

$$h^T \left(\nabla^2 f(\bar{x}) + \sum_{i=1}^{m+k} \bar{y}_i \nabla^2 g_i(\bar{x}) \right) h \leq 0$$

and this is a contradiction to iv).

Example (WOLFE [78]): For $n \geq 2$ and $k = 2,3,4$ define $s_k : \mathbb{R}^n \to \mathbb{R}$ by

$$s_k(x) := \sum_{j=1}^{n} x_j^k.$$

Consider the problem

(P) Minimize $f(x) := s_3(x)^2 - s_2(x)s_4(x)$ on

$$M := \{x \in \mathbb{R}^n : 0 \leq x_j \leq 1 \ (j=1,\dots,n)\}.$$

We shall show that $\bar{x} \in \mathbb{R}^n$ gives a local minimum of (P) if m components of $\bar{x}$ are 1 and p components are 1/2, where $m + p = n$ and $n > m > n/9$. We apply the sufficient conditions of second order given in Theorem 5.5.4. These say: $\bar{x} \in \mathbb{R}^n$ is a strict local solution of (P) if there exist $\lambda, \mu \in \mathbb{R}^n$ with

i) $\lambda, \mu \geq 0$

ii) $\lambda^T \bar{x} = 0$, $\mu^T(e - \bar{x}) = 0$ with $e = (1,\dots,1)^T \in \mathbb{R}^n$

iii) $\nabla f(\bar{x}) - \lambda + \mu = 0$ resp.

$$6\bar{x}_i^2 \sum_{j=1}^{n} \bar{x}_j^3 - 2\bar{x}_i \sum_{j=1}^{n} \bar{x}_j^4 - 4\bar{x}_i^3 \sum_{j=1}^{n} \bar{x}_j^2 - \lambda_i + \mu_i = 0 \qquad (i = 1,\ldots, n)$$

iv) $h^T\nabla^2 f(\bar{x})h > 0$ for all $h \in L' = \{h \in \mathbb{R}^n : h_i = 0$ if $\lambda_i > 0$ or $\mu_i > 0,\ i = 1,\ldots,n\},\ h \neq 0$.

Let $I_1, I_2 \subset \{1,\ldots,n\}$ be nonempty and disjoint, $I_1 \cup I_2 = \{1,\ldots,n\}$. Further let $n > |I_1| =: m > n/9$ and $p := |I_2|$ $(= n-m)$. Define $\bar{x} \in \mathbb{R}^n$ by

$$\bar{x}_i = \begin{cases} 1 & \text{for } i \in I_1 \\ 1/2 & \text{for } i \in I_2. \end{cases}$$

Further let $\lambda = 0$ and

$$\mu_i = \begin{cases} 3p/8 & \text{for } i \in I_1 \\ 0 & \text{for } i \in I_2. \end{cases}$$

Then i), ii), iii) are satisfied and it remains to show that

$$h^T\nabla^2 f(\bar{x})h > 0 \text{ for } h \in L' = \{h \in \mathbb{R}^n : h_i = 0 \text{ for } i \in I_1\}, h \neq 0.$$

By an easy calculation one has

$$\frac{\partial^2 f}{\partial x_i \partial x_j}(\bar{x}) = \begin{cases} 1/8 & \text{for } i,j \in I_2,\ i \neq j \\ 1/8 + m - p/8 & \text{for } i = j \in I_2. \end{cases}$$

For $h \in L'$ then

$$\begin{aligned} h^T\nabla^2 f(\bar{x})h &= \sum_{i,j\in I_2} \frac{\partial^2 f}{\partial x_i \partial x_j}(\bar{x}) h_i h_j \\ &= (1/8+m-p/8) \sum_{i\in I_2} h_i^2 + 1/8 \sum_{\substack{i,j\in I_2 \\ i\neq j}} h_i h_j \\ &= (m-p/8) \sum_{i\in I_2} h_i^2 + 1/8 \Big(\sum_{i\in I_2} h_i\Big)^2 \end{aligned}$$

$$= \frac{9m-n}{8} \sum_{i\in I_2} h_i^2 + \frac{1}{8}\left(\sum_{i\in I_2} h_i\right)^2$$

$$> 0 \text{ if } h \neq 0, \text{ since } m > n/9,$$

so the $\overline{x}$ above is a strict local solution of (P) and one can show (we will not do so, however) that all solutions are of this form.

5.6 Literature

For 5.1: An introduction to the differential calculus in normed linear spaces is contained in several texts on optimization, e.g. LUENBERGER [53], IOFFE-TICHOMIROV [37].

For 5.2: The fundamental paper on the Theorem of LYUSTERNIK is LYUSTERNIK [55]; c.f. also LYUSTERNIK-SOBOLEV [56]. Theorem 5.2.3, the generalization of the open mapping theorem, is proved in ZOWE-KURCYUSZ [81]. Lemma 3.1 of that paper also gives the main result of this section, namely Theorem 5.2.5, and refers the reader for the proof to ROBINSON [67]. However ROBINSON uses the theory of convex processes, so the proof given here is surely easier. A similar "elementary" proof of the Theorem of LYUSTERNIK is also given in BROKATE-KRABNER [10]. In this connection one should also mention LEMPIO [51].

For 5.3: The literature on necessary optimality conditions of first order resp. Lagrange multiplier rules of KUHN-TUCKER and F. JOHN type is extremely extensive. Among the historical papers we shall only mention besides that of LYUSTERNIK [55] for optimization problems with equations as constraints those of KARUSH [40], JOHN [38] and KUHN-TUCKER [48] for finite dimensional programs with inequalities as side conditions. A thorough exposition of necessary optimality conditions in finite dimensional optimization problems can also be found in MANGASARIAN [57], HESTENES [33]; BAZARAA-SHETTY [3]. A small selection of literature on necessary optimality conditions of first order for not necessarily finite dimensional problems is given by LUENBERGER [53], IOFFE-TICHOMIROV [37], PONSTEIN [65], KIRSCH-WARTH-WERNER [42], GIRSANOV [27], NEUSTADT [62], COLONIUS [15], BEN TAL-ZOWE [4], PENOT [63].

For 5.4: For applications of abstract necessary optimality conditions to problems of the calculus of variations and of control theory we used LUENBERGER [53], IOFFE-TICHOMIROV [37] and KIRSCH-WARTH-WERNER [42]. How one gets from the local to the global Pontryagin maximum principle can be found in GIRSANOV [27].

For 5.5: Theorem 5.5.2 is due to MAURER-ZOWE [59]. Other important papers in this connection are e.g. HOFFMANN-KORNSTAEDT [35], BEN TAL-ZOWE [4], LINNEMANN [52]. Necessary and sufficient optimality conditions of second order for finite dimensional optimization problems are contained in FIACCO-McCORMICK [22], LUENBERGER [54].

§ 6 EXISTENCE THEOREMS FOR SOLUTIONS OF OPTIMIZATION PROBLEMS

6.1 Funktional analytic existence theorems

In this paragraph we consider optimization problems

(P) Minimize $f(x)$ on $M \subset X$,

assume that $(X, \|\ \|)$ is a normed linear space and want to show the existence of a global solution $\bar{x}$ for (P) under appropriate hypotheses on the objective function $f : X \to \mathbb{R}$ and the set of feasible solutions $M \subset X$. The idea behind many existence proofs for (P) is the following:

1. Choose $x_o \in M$, define $M_o := \{x \in M : f(x) \leq f(x_o)\}$ and consider the equivalent problem

   (P_o) Minimize $f(x)$ on M_o.

2. Show that M_o is "compact" and f is "continuous" and that a "continuous" real-valued function on a "compact" set assumes its minimum, i.e. that for suitably defined compactness and continuity a theorem of Weierstrass type holds.

We shall illustrate this method with an example.

Example: Suppose $(X, \|\ \|)$ is a normed linear space. A function $f : X \to \mathbb{R}$ is uniformly convex (on X) if there is a constant $c > 0$ with

$$x,y \in X,\ \lambda \in [0,1] \Rightarrow$$

$$\frac{c}{2}\,\lambda(1-\lambda)\,\| x-y \|^2 \leq (1-\lambda)f(x)+\lambda f(y)-f((1-\lambda)x+\lambda y).$$

We wish to show that: if $f : \mathbb{R}^n \to \mathbb{R}$ is uniformly convex, then there exists exactly one $\bar{x} \in \mathbb{R}^n$ with

$$f(\bar{x}) = \inf\,\{f(x) : x \in \mathbb{R}^n\}.$$

For: Since f is convex on all of $\mathbb{R}^n$ it is also continuous

(Corollary 3.3.9). Let $x_o \in \mathbb{R}^n$ be arbitrary and

$$M_o := \{x \in \mathbb{R}^n : f(x) \leq f(x_o)\}.$$

The existence of an $\bar{x} \in \mathbb{R}^n$ with $f(\bar{x}) = \inf \{f(x) : x \in \mathbb{R}^n\}$ is proven if we can show that M_o is bounded.

By Theorem 3.3.6 there exists an $l_o \in \partial f(x_o)$, that is, there is an $l_o \in \mathbb{R}^n$ with $l_o^T(x-x_o) \leq f(x) - f(x_o)$ for all $x \in \mathbb{R}^n$.

Let $x \in M_o$. By the uniform convexity there is a $c > 0$ with:

$$\begin{aligned}
\frac{1}{8} c|x-x_o|^2 &\leq \frac{1}{2} (f(x)+f(x_o)) - f(\tfrac{1}{2}(x+x_o)) \\
&= \frac{1}{2} (f(x)-f(x_o)) - (f(\tfrac{1}{2}(x+x_o))-f(x_o)) \\
&\leq - (f(\tfrac{1}{2}(x+x_o))-f(x_o)) \\
&\leq - \frac{1}{2} l_o^T(x-x_o) \\
&\leq \frac{1}{2} |l_o||x-x_o|
\end{aligned}$$

and thus $M_o \subset B[x_o;4|l_o|/c]$ is bounded. f has exactly one minimum by uniform convexity.

In this example we used the fact that a continuous function on a compact subset of $\mathbb{R}^n$ takes on its extrema, in particular its minimum. It is well known that this is true more generally:

<u>6.1.1 Theorem:</u> Suppose $(X,\|\ \|)$ is a normed linear space, $M_o \subset X$ compact (i.e. for every sequence $\{x_k\} \subset M_o$ there is a subsequence $\{x_{k_i}\} \subset \{x_k\}$ converging to some $x \in M_o$) and $f : M_o \to \mathbb{R}$ is lower semicontinuous (i.e. if $x_o \in M_o$ and $\{x_k\} \subset M_o$ converges to x_o, then

$$f(x_o) \leq \liminf_{k\to\infty} f(x_k)).$$

Then there exists an $\bar{x} \in M_o$ with $f(\bar{x}) = \inf \{f(x) : x \in M_o\}$.

Proof: 1) $\inf \{f(x) : x \in M_o\} > -\infty$.
For: If we had $\inf \{f(x) : x \in M_o\} = -\infty$, then there would exist a sequence $\{x_k\} \subset M_o$ with $f(x_k) \leq -k$ for all $k \in \mathbb{N}$. Since M_o is compact we can choose a convergent subsequence $\{x_{k_i}\} \subset \{x_k\}$ converging to an $x \in M_o$. Since f is lower semi-continuous we have

$$-\infty < f(x) \leq \liminf_{i\to\infty} f(x_{k_i}).$$

This is a contradiction to the fact that $f(x_{k_i}) \leq -k_i$.

2) Let $\{x_k\} \subset M_o$ be a sequence with

$$\lim_{k\to\infty} f(x_k) = \inf \{f(x) : x \in M_o\}.$$

Again we can choose a convergent subsequence $\{x_{k_i}\}$ converging to some $\bar{x} \in M_o$. But then

$$\begin{aligned} \inf \{f(x) : x \in M_o\} \leq f(\bar{x}) &\leq \liminf_{i\to\infty} f(x_{k_i}) \\ &= \lim_{i\to\infty} f(x_{k_i}) \\ &= \inf \{f(x) : x \in M_o\} \end{aligned}$$

from which the claim follows.

In infinite dimensional normed linear spaces one can seldom apply Theorem 6.1.1 - roughly speaking for the reason that there are too few compact sets there (with respect to the norm topology). (For example the unit ball in a normed linear space is compact if and only if the space is finite dimensional.) This is precisely the reason why the weak topology in a normed linear space is so important. We want to present the functional analytic tools needed here on as elementary a level as possible, so we will not define the weak topology but only the concept of weak convergence.

6.1.2 Definition: Suppose $(X, \|\ \|)$ is a normed linear space. A sequence $\{x_k\} \subset X$ is weakly convergent to an $x \in X$

$$(w - \lim_{k\to\infty} x_k = x \text{ or } x_k \rightharpoonup x),$$

if

$$\lim_{k\to\infty} \langle x^*, x_k\rangle = \langle x^*, x\rangle \text{ for all } x^* \in X^*.$$

Remarks: 1. One has the usual rules:

$$x_k \rightharpoonup x,\ y_k \rightharpoonup y,\ \alpha,\beta \in \mathbb{R} \Rightarrow \alpha x_k + \beta y_k \rightharpoonup \alpha x + \beta y.$$

2. The weak limit of a weakly convergent sequence is uniquely determined. For if $\langle x^*, z\rangle = 0$ for all $x^* \in X^*$ then $z = 0$, since

$$\| z \| = \max_{\| x^* \| \le 1} \langle x^*, z\rangle \quad \text{(Corollary 3.2.7).}$$

3. If $\{x_k\}$ is (strongly) convergent to $x \in X$, i.e.

$$\| x_k - x \| \to 0,$$

then $\{x_k\}$ also converges weakly to x. If X is finite dimensional, then the converse also holds (proof?).

Now building on the concept of weak convergence we can define:

6.1.3 Definition: Suppose $(X, \| \ \|)$ is a normed linear space.

i) A subset $A \subset X$ is weakly sequentially closed if

$$\{x_k\} \subset A,\ x_k \rightharpoonup x \Rightarrow x \in A.$$

ii) A subset $A \subset X$ is weakly sequentially compact if for every sequence $\{x_k\} \subset A$ there exists a subsequence $\{x_{k_i}\} \subset \{x_k\}$ and an $x \in A$ with $x_{k_i} \rightharpoonup x$.

iii) Suppose $A \subset X$ and $f : A \to \mathbb{R}$. f is weakly sequentially lower semicontinuous on A if

$$\{x_k\} \subset A,\ x_k \rightharpoonup x \in A \Rightarrow f(x) \le \liminf_{k\to\infty} f(x_k).$$

Remark: Suppose $(X, \|\ \ \|)$ is a normed linear space and $A \subset X$. Obviously one has:

1. A weakly sequentially closed $\Rightarrow$ A closed.

2. A compact $\Rightarrow$ A weakly sequentially compact.

3. Suppose $f : A \to \mathbb{R}$. Then f weakly sequentially lower semicontinuous $\Rightarrow$ f lower semicontinuous.

If X is finite dimensional then the corresponding converses also hold.

Analogous to Theorem 6.1.1 one now has (the proof is almost identical: one only has to replace the strong convergence by weak convergence):

6.1.4 Theorem: Suppose $(X, \|\ \ \|)$ is a normed linear space, $M_o \subset X$ weakly sequentially compact and $f : M_o \to \mathbb{R}$ weakly sequentially lower semicontinuous on M_o. Then there exists an $\bar{x} \in M_o$ with

$$f(\bar{x}) = \inf \{f(x) : x \in M_o\}.$$

In order to apply Theorem 6.1.4 we need necessary conditions for M_o to be weakly sequentially compact and f to be weakly sequentially lower semicontinuous on M_o.

Since a weakly sequentially compact set is obviously necessarily weakly sequentially closed, the following result will prove useful.

6.1.5 Theorem: Suppose $(X, \|\ \ \|)$ is a normed linear space. If $A \subset X$ is nonempty, closed and convex, then A is weakly sequentially closed.

Proof: Suppose $\{x_k\} \subset A, x_k \rightharpoonup x$. Assume that $x \notin A$. By the strict separation theorem 3.2.5 there exists an $x^* \in X^*$ and a $\gamma \in \mathbb{R}$ with $<x^*,x> < \gamma < <x^*,a>$ for all $a \in A$. Since $\{x_k\} \subset A$, we have $<x^*,x> < \gamma < <x^*,x_k>$ for all $k \in \mathbb{N}$. Since $x_k \rightharpoonup x$ it follows that $<x^*,x> < \gamma \le <x^*,x>$, a contradiction.

An application of this theorem gives a sufficient condition

for $f : M \to \mathbb{R}$ to be weakly sequentially lower semicontinuous.

<u>6.1.6 Theorem</u>: Suppose $(X, \| \ \|)$ is a normed linear space, $M \subset X$ is nonempty, convex and closed and $f : M \to \mathbb{R}$ is continuous and convex. Then f is weakly sequentially lower semicontinuous on M.

<u>Proof</u>: Suppose $\{x_k\} \subset M$, $x_k \rightharpoonup x$. Assume that

$$\liminf_{k\to\infty} f(x_k) < f(x).$$

Let $\lambda \in (\liminf_{k\to\infty} f(x_k), f(x))$ and $M_\lambda := \{y \in M : f(y) \leq \lambda\}$.

By the convexity of M and of f we have that M_λ is convex; by the closedness of M and the continuity of f we have that M_λ is closed. Furthermore there exists a subsequence $\{x_{k_i}\} \subset \{x_k\}$ with $\{x_{k_i}\} \subset M_\lambda$ and since the entire sequence $\{x_k\}$ weakly converges to x it follows that $x_{k_i} \rightharpoonup x$. By Theorem 6.1.5 it follows M_λ is weakly sequentially closed and thus $x \in M_\lambda$ resp. $f(x) \leq \lambda$, a contradiction to $\lambda < f(x)$.

<u>Example</u>: Suppose $(X, \| \ \|)$ is a normed linear space, $z \in X$ and $f(x) := \| x-z \|$. Then f is convex and continuous (e.g. $|f(x)-f(y)| \leq \| x-y \|$ for all $x,y \in X$) and thus by Theorem 6.1.6 weakly sequentially lower semicontinuous.

We still need a sufficient condition for a subset M of a normed linear space X to be weakly sequentially compact.

<u>Example</u>: If one gives $X = C[a,b]$ the norm

$$\| x \| = \max_{t\in[a,b]} |x(t)|,$$

then the closed unit ball $B[0;1]$ is not weakly sequentially compact (proof?).

Because of this example one cannot expect that the closed unit ball in any normed linear space will be weakly sequentially compact. On the other hand this assertion is true in many spaces of practical importance. To see this we need:

6.1.7 Definition: A normed linear space $(X, \| \ \|)$ is *reflexive* if the map $i : X \to X^{**} := (X^*)^*$ defined by $i(x)x^* := \langle x^*, x\rangle$ is surjective.

Remarks: 1) Suppose $(X, \| \ \|)$ is a normed linear space and $i : X \to X^{**}$ is given by $i(x)x^* = \langle x^*, x\rangle$. Then one has

i) i is linear.

ii) $\| i(x) \| = \| x \|$ for all $x \in X$. Proof: For arbitrary $x^* \in X^*$ one has

$$|i(x)x^*| \leq \| x \| \| x^* \|$$

and thus $\| i(x) \| \leq \| x \|$. On the other hand by Corollary 3.2.7 for every $x \in X$ there is an $x^* \in X^*$ with $\| x^* \| \leq 1$ and $\| x \| = \langle x^*, x\rangle = |i(x)x^*| \leq \| i(x) \|$.

Thus by i), ii) one can always identify X with the linear subspace $i(X)$ in X^{**}. If X is reflexive, then one can identify X with X^{**}.

2) Since the dual space of a normed linear space is complete, a reflexive normed linear space is necessarily a Banach space.

Examples: 1) Every finite dimensional normed linear space is reflexive.

2) Suppose X is a (real) linear space and $(\cdot,\cdot) : X \times X \to \mathbb{R}$ a map (inner product) with

a) $(x,x) \geq 0$ for all $x \in X$, $(x,x) = 0 \iff x = 0$

b) $(x,y) = (y,x)$ for all $x,y \in X$

c) $(\alpha x,y) = \alpha(x,y)$ for all $\alpha \in \mathbb{R}$, $x,y \in X$

d) $(x+y,z) = (x,z) + (y,z)$ for all $x,y,z \in X$.

Then one says $(X, (\ , \ ))$ is a (real) *pre-Hilbert space*. One defines a norm by

$$\| x \| := (x,x)^{1/2}$$

(the triangle inequality is proved using the CAUCHY-SCHWARZ inequality

$$|(x,y)|^2 \leq (x,x)(y,y) \text{ for all } x,y \in X).$$

If the normed pre-Hilbert space $(X,(\ ,\ ))$ so defined is complete, i.e. every Cauchy sequence convergent, then $(X,(\ ,\ ))$ is a (real) <u>Hilbert space</u>. The most important examples of Hilbert spaces are $X = \mathbb{R}^n$ with $(x,y) = x^T y$ and $X = L^2[a,b]$ with

$$(x,y) = \int_a^b x(t)y(t)dt.$$

Every Hilbert space is reflexive. This comes from the fact that one can identify a (real) Hilbert space X with its dual space X^*. This is essentially the assertion of the RIESZ representation theorem, which we shall state without proof:

Suppose $(X,(\ ,\ ))$ is a (real) Hilbert space and $j : X \to X^*$ is defined by $j(x)y = (x,y)$. Then j is linear, isometric (i.e. $\| j(x) \| = \| x \|$ for all $x \in X$) and surjective.

If one defines $(\ ,\ ) : X^* \times X^* \to \mathbb{R}$ by

$$(x^*,y^*) = (j^{-1}(x^*),j^{-1}(y^*)),$$

then $(X^*,(\ ,\ ))$ is also a Hilbert space.

The following theorem will not be proven here. We refer the reader to YOSIDA [79, p. 126] for the proof.

<u>6.1.8 Theorem (EBERLEIN-SHMULYAN)</u>: A Banach space $(X,\|\ \|)$ is reflexive if and only if every (strongly) bounded sequence $\{x_k\}$ (i.e. there exists a constant $C > 0$ with $\| x_k \| \leq C$ for all $k \in \mathbb{N}$) has a subsequence $\{x_{k_i}\} \subset \{x_k\}$ converging weakly to an element $x \in X$.

As a corollary to 6.1.5 and 6.1.8 we have:

<u>6.1.9 Corollary</u>: A nonempty, convex, closed and bounded subset of a reflexive Banach space is weakly sequentially compact.

Further by Theorems 6.1.4, 6.1.6 and Corollary 6.1.9 we have:

<u>6.1.10 Theorem</u>: Suppose given the optimization problem

(P) Minimize $f(x)$ on $M \subset X$.

Here $(X, \|\ \|)$ shall be a reflexive Banach space, $M \subset X$ nonempty, closed and convex and $f : M \to \mathbb{R}$ convex and continuous. For some $x_o \in M$ suppose that the level set

$$M_o := \{x \in M : f(x) \leq f(x_o)\}$$

is bounded. Then (P) has a solution $\bar{x} \in M$.

The following special case of this theorem has to do with the solvability of approximation problems.

<u>6.1.11 Theorem</u>: Suppose $(X, \|\ \|)$ is a reflexive Banach space, $z \in X$ and $C \subset X$ nonempty, convex and closed. Then there exists an $\bar{x} \in C$ with

$$\| \bar{x}-z \| = \inf_{c \in C} \| c-z \|.$$

<u>Proof</u>: Let $f(x) := \| x-z \|$, $M := C$ and apply Theorem 6.1.10.

<u>Remarks</u>: 1) If the X in Theorem 6.1.11 is even <u>strictly normed</u>, i.e. if

$$x, y \in X \setminus \{0\}, \| x+y \| = \| x \| + \| y \| \Rightarrow \exists\ \lambda > 0 \text{ with } x = \lambda y,$$

then there is precisely one $\bar{x} \in C$ with

$$\| \bar{x}-z \| = \inf_{c \in C} \| c-z \| \quad \text{(proof?)}$$

2) A Hilbert space $(X, (\ ,\ ))$ (c.f. Example 2) following Defi-

nition 6.1.7) is reflexive and strictly normed (proof?). The approximation problem with respect to a nonempty, convex and closed subset $C \subset X$ is thus uniquely solvable (which one can of course prove more directly, c.f. e.g. LUENBERGER [53, p. 69]). If $\bar{x} \in C$ is the solution, then

$$0 \le \lim_{t \to 0+} \frac{\| \bar{x}+t(c-\bar{x})-z \|^2 - \| \bar{x}-z \|^2}{t} = 2(\bar{x}-z, c-\bar{x})$$

i.e. $(z-\bar{x}, c-\bar{x}) \le 0$ for all $c \in C$ (geometric interpretation?). This necessary optimality condition is also sufficient (proof?).

6.2 Existence of optimal controls

In the previous section we gave several variations on the theorem of Weierstrass which says that a real-valued function on a compact set takes on its minimum. We shall now give an example for the application of these functional analytic existence theorems.

We shall demonstrate the existence of a solution of a problem in optimal control theory. We consider the following problem:

A linear process described by the linear system of differential equations

$$\dot{x} = A(t)x + B(t)u$$

is to be controlled during a time intervall $[t_o, t_1]$ starting in an initial state $x(t_o) = x_o$ and ending in a terminal state $x(t_1) \in Q_1$ such that on the one hand $u(t) \in \Omega$ for a.a. $t \in [t_o, t_1]$ and on the other hand the objective function

$$I(x,u) = g(x(t_1)) + \int_{t_o}^{t_1} \{f^o(x(t),t) + h^o(u(t),t)\} dt$$

is minimal.

The following assumptions are made:

(A) i) $A(\cdot)$, $B(\cdot)$ are continuous $n \times n$ resp. $n \times m$ matrices on $[t_o, t_1]$.

ii) $x_o \in \mathbb{R}^n$, $Q_1 \subset \mathbb{R}^n$ is convex and closed.

iii) $\Omega \subset \mathbb{R}^m$ is convex and closed.

iv) $g : \mathbb{R}^n \to \mathbb{R}$, $f^o : \mathbb{R}^n \times [t_o,t_1] \to \mathbb{R}$ and $h^o : \mathbb{R}^m \times [t_o,t_1] \to \mathbb{R}$ are continuous. For every $t \in [t_o,t_1]$ the function $h(\cdot,t) : \mathbb{R}^m \to \mathbb{R}$ is convex.

v) There exist $b,c \in \mathbb{R}$ with $g(x) \geq b + c|x|$ for all $x \in \mathbb{R}^n$ and $f^o(x,t) \geq b + c|x|$ for all $(x,t) \in \mathbb{R}^n \times [t_o,t_1]$. Without loss of generality we may assume $c \leq 0$.

vi) There exists a constant $a > 0$ with $h^o(u,t) \geq a|u|^2$ for all $(u,t) \in \mathbb{R}^m \times [t_o,t_1]$.

Important special cases are: $Q_1 = \{x_1\}$ (fixed terminal state), $Q_1 = \mathbb{R}^n$ (free terminal state), $\Omega \subset \mathbb{R}^m$ compact (e.g. the unit ball) or $\Omega = \mathbb{R}^m$ (no control condition), $f^o(x,t) = \frac{1}{2} x^T Q(t)x$, $h^o(u,t) = \frac{1}{2} u^T R(t)u$ with continuous, positive semidefinite resp. positive definite $n \times n$ resp. $m \times m$ matrices $Q(\cdot)$ resp. $R(\cdot)$, g bounded below or convex (e.g. $g(x) \geq g(0) - |l_o||x|$ for $l_o \in \partial g(0)$).

In order to formulate this problem in optimal control theory somewhat more simply we define a map

$$S : L^2_m[t_o,t_1] \to C_n[t_o,t_1]$$

by: $x(t) = (Su)(t)$ is the solution of the initial value problem

$$\dot{x} = A(t)x + B(t)u(t), \; x(t_o) = x_o,$$

i.e.

$$Su(t) := \Phi(t)x_o + \Phi(t) \int_{t_o}^{t} \Phi^{-1}(s)B(s)u(s)ds,$$

where $\Phi(\cdot)$ is the fundamental system for $\dot{x} = A(t)x$ normalized by $\Phi(t_o) = I$. Then the problem reads

$$\text{(P)}\quad \text{Minimize } J(u) := g(Su(t_1)) + \int_{t_o}^{t_1} \{f^o(Su(t),t)+h^o(u(t),t)\}dt$$

$$\text{on } M := \{u \in L_m^2[t_o,t_1] : Su(t_1) \in Q_1, u(t) \in \Omega \text{ a.e. on } [t_o,t_1]\}.$$

Now we show:

If the hypotheses (A) i) - vi) are satisfied and if there is a $u_o \in M$ with $J(u_o) < \infty$, then (P) has a solution $\bar{u} \in M$.

To prove this we let $X := L_m^2[t_o,t_1]$ and provide X with the inner product

$$(u,v) = \int_{t_o}^{t_1} u(t)^T v(t)dt$$

and the associated norm

$$\| u \|_2 := \left(\int_{t_o}^{t_1} |u(t)|^2 dt\right)^{1/2}.$$

Then $(L_m^2[t_o,t_1],(\ ,\ ))$ is a Hilbert space and hence $(L_m^2[t_o,t_1],\|\ \ \|_2)$ is a reflexive Banach space. We wish to apply Theorem 6.1.4 with $M_o := \{u \in M : J(u) \le J(u_o)\}$, $f = J$ and so we demonstrate that

1) M_o is bounded

2) M_o is weakly sequentially closed.

(From 1) and 2) it follows that M_o is weakly sequentially compact.)

3) $J : M_o \to \mathbb{R}$ is weakly sequentially lower semicontinuous on M_o.

For $u,v \in L_m^2[t_o,t_1]$ and $t \in [t_o,t_1]$ we have

$$\begin{aligned}
|Su(t)-Sv(t)| &\le |\Phi(t)| \int_{t_o}^{t} |\Phi^{-1}(s)B(s)|\,|u(s)-v(s)|ds \\
&\le |\Phi(t)| \left(\int_{t_o}^{t_1} |\Phi^{-1}(s)B(s)|^2 ds\right)^{1/2} \| u-v \|_2 \\
&\le c_1 \| u-v \|_2 \text{ for a constant } c_1 > 0,
\end{aligned}$$

and hence $\| Su-Sv \|_\infty \le c_1 \| u-v \|_2$. Thus

$$S : (L_m^2[t_o,t_1], \| \quad \|_2) \to (C_n[t_o,t_1], \| \quad \|_\infty)$$

is continuous and there exist constants $c_o, c_1 > 0$ with

$$\| Su \|_\infty \le c_o + c_1 \| u \|_2 \text{ for all } u \in L_m^2[t_o,t_1].$$

1): Suppose $u \in M_o$. Then we have

$$J(u_o) \ge J(u) = g(Su(t_1)) + \int_{t_o}^{t_1} \{f^o(Su(t),t)+h^o(u(t),t)\}dt$$

$$\ge b + c|Su(t_1)| + \int_{t_o}^{t_1} (b+c|Su(t)|)dt + a\| u \|_2^2$$

$$\ge \tilde{b} + \tilde{c}\| u \|_2 + a\| u \|_2^2$$

with constants $\tilde{b}, \tilde{c}$. Since $a > 0$ the boundedness of M_o follows.

2): Suppose $\{u_k\} \subset M_o$, $u_k \rightharpoonup u$. For every $v \in L_m^2[t_o,t_1]$ one has

$$\int_{t_o}^{t_1} v(t)^T u_k(t)dt \to \int_{t_o}^{t_1} v(t)^T u(t)dt$$

(RIESZ representation theorem!). In particular

$$Su_k(t) \to Su(t) \text{ for all } t \in [t_o,t_1].$$

a) $u \in M$. We show more than that: M is convex and closed and thus by Theorem 6.1.5 weakly sequentially closed. The convexity of

$$M := \{u \in L_m^2[t_o,t_1] : Su(t_1) \in Q_1, u(t) \in \Omega \text{ a.e. on } [t_o,t_1]\}$$

follows from the convexity of Q_1 and Ω ((A) ii), iii)) and from the fact that S is an affine linear map.

Suppose $\{u_k\} \subset M$ and $u_k \to u$, i.e. $\| u_k-u \|_2 \to 0$. Then $\{Su_k(t_1)\} \subset Q_1$ converges to $Su(t_1)$, and since Q_1 is closed we have $Su(t_1) \in Q_1$. Furthermore we can choose a subsequence

$\{u_{k_i}\} \subset \{u_k\}$ converging to u pointwise a.e. on $[t_o,t_1]$. This is often proved in connection with the completeness of $L^2[t_o,t_1]$ (c.f. e.g. YOSIDA [79, p. 53]). Since Ω is closed ((A) iii)), $u(t) \in \Omega$ for a.a. $t \in [t_o,t_1]$. Thus altogether the closedness of M is proved.

b) $J(u) \leq J(u_o)$.

We have $u_k \rightharpoonup u$ and thus $Su_k(t) \to Su(t)$ for all $t \in [t_o,t_1]$. We show that $\{Su_k\} \subset C_n[t_o,t_1]$ contains a subsequence $\{Su_{k_i}\}$ which even converges uniformly to Su on $[t_o,t_1]$. This we do with the help of the theorem of ARZELA-ASCOLI (c.f. e.g. YOSIDA [79, p. 85]):

if $\{x_k\} \subset C_n[t_o,t_1]$ is bounded (i.e. there exists a constant $d > 0$ with $\| x_k \|_\infty \leq d$ for all $k \in \mathbb{N}$) and equicontinuous (i.e. for every $\varepsilon > 0$ there exists a $\delta = \delta(\varepsilon) > 0$ with $|x_k(t)-x_k(s)| \leq \varepsilon$ for all $s,t \in [t_o,t_1]$ with $|t-s| \leq \delta$ and all $k \in \mathbb{N}$), then one can choose a uniformly convergent subsequence $\{x_{k_i}\} \subset \{x_k\}$

We have $\| Su_k \|_\infty \leq c_o + c_1 \| u_k \|_2$. Since M_o is bounded (c.f. 1)) and $\{u_k\} \subset M_o$ we also have that $\{\| Su_k \|_\infty\}$ is bounded.

For $s,t \in [t_o,t_1]$ one has

$$Su_k(t) - Su_k(s) = \int_s^t (A(\tau)Su_k(\tau) + B(\tau)u_k(\tau))d\tau .$$

From the boundedness of $\{Su_k\} \subset C_n[t_o,t_1]$ and $\{u_k\} \subset L^2_m[t_o,t_1]$ we get the existence of constants $d_o, d_1 > 0$ with

$$|Su_k(t)-Su_k(s)| \leq d_o|t-s| + d_1|t-s|^{1/2}.$$

Thus $\{Su_k\}$ is also equicontinuous.

In the proof that $J(u) \leq J(u_o)$ we can therefore without loss of generality assume that $u_k \rightharpoonup u$ and $Su_k \to Su$ uniformly on $[t_o,t_1]$ and naturally $J(u_k) \leq J(u_o)$.

One has

$$J(u_o) - J(u) = J(u_o) - J(u_k) + J(u_k) - J(u)$$

$$\geq J(u_k) - J(u)$$

$$= g(Su_k(t_1)) - g(Su(t_1))$$

$$+ \int_{t_o}^{t_1} (f^o(Su_k(t),t) - f^o(Su(t),t))dt$$

$$+ \int_{t_o}^{t_1} h^o(u_k(t),t)dt - \int_{t_o}^{t_1} h^o(u(t),t)dt$$

and thus

$$J(u_o) - J(u) \geq \liminf_{k\to\infty} \int_{t_o}^{t_1} h^o(u_k(t),t)dt - \int_{t_o}^{t_1} h^o(u(t),t)dt.$$

We now show that

$$\liminf_{k\to\infty} \int_{t_o}^{t_1} h^o(u_k(t),t)dt \geq \int_{t_o}^{t_1} h^o(u(t),t)dt.$$

Assume this were not the case. Then there would exist a constant λ with

$$\liminf_{k\to\infty} \int_{t_o}^{t_1} h^o(u_k(t),t)dt < \lambda < \int_{t_o}^{t_1} h^o(u(t),t)dt.$$

Thus there exists a subsequence $\{u_{k_i}\} \subset \{u_k\}$ with

$$\{u_{k_i}\} \subset A_\lambda := \{v \in L_m^2[t_o,t_1] : \int_{t_o}^{t_1} h^o(v(t),t)dt \leq \lambda\}.$$

By the convexity of $h^o(\cdot,t) : \mathbb{R}^m \to \mathbb{R}$ it follows A_λ is convex. We show that A_λ is closed. Then by Theorem 6.1.5 it follows that A is also weakly sequentially closed and from $u_{k_i} \rightharpoonup u$ it follows $u \in A$, a contradiction. Except for the closedness of A the assertion $J(u) \leq J(u_o)$ is proved. Suppose $\{v_k\} \subset A_\lambda$ and $v_k \to v$ (i.e. $\| v_k - v \|_2 \to 0$). As in a) we deduce that there is a subsequence $\{v_{k_i}\} \subset \{v_k\}$ converging pointwise to v a.e. on $[t_o, t_1]$. Hence we have

1. $\quad 0 \le h^o(v_{k_i}(t), t)$

2. $\quad \lim_{i\to\infty} h^o(v_{k_i}(t), t) = h^o(v(t), t)$ a.e. on $[t_o, t_1]$.

3. $\quad \int_{t_o}^{t_1} h^o(v_{k_i}(t), t)dt \le \lambda.$

By FATOU's lemma it follows

$$\int_{t_o}^{t_1} h^o(v(t), t)dt \le \liminf_{i\to\infty} \int_{t_o}^{t_1} h^o(v_{k_i}(t), t)dt \le \lambda$$

resp. $v \in A_\lambda$. Thus A_λ is closed.
And thus 2) is finally proved.

3) Now the proof that $J : M_o \to \mathbb{R}$ is weakly sequentially lower semicontinuous on M_o is easy. Suppose $\{u_k\} \subset M_o$, $u_k \rightharpoonup u$. Assume

$$\liminf_{k\to\infty} J(u_k) < \lambda < J(u).$$

Then there exists a subsequence $\{u_{k_i}\} \subset \{u_k\}$ with $J(u_{k_i}) \le \lambda$. Without loss of generality one may assume that $\{Su_{k_i}\} \subset C_n[t_o, t_1]$ converges uniformly to Su. As in 2) b) it then follows from $J(u_{k_i}) \le \lambda$ that also $J(u) \le \lambda$ and that is a contradiction. Therefore

$$J(u) \le \liminf_{k\to\infty} J(u_k),$$

so 3) is also proved and altogether we have demonstrated the existence of a solution of the problem in optimal control (P) under the hypotheses (A) i) - vi).

6.3 Literature

For 6.1: More explicit discussions on weak convergence, weak topology, reflexive Banach spaces etc. can be found in many books about functional analysis, e.g. YOSIDA [79].

For 6.2: Similar existence statements are proved in LEE-MARKUS [49]. For more general statements the reader is referred to e.g. FLEMING-RISHEL [24].

BIBLIOGRAPHY

[1] APOSTOL, T.M.: Mathematical Analysis. A Modern Approach to Advanced Calculus. Reading, Mass.: Addison-Wesley, 1957

[2] ARROW, K.J.; KARLIN, S.: Production over time with increasing marginal costs. In: Studies in the Mathematical Theory of Inventory and Production. Eds.: Arrow, K.J.; Karlin, S.; Scarf, H. Stanford, Calif.: Stanford University Press, 1958

[3] BAZARAA, M.S.; SHETTY, C.M.: Foundations of Optimization. Lecture Notes in Economics and Mathematical Systems 122. Berlin-Heidelberg-New York: Springer, 1976

[4] BEN TAL, A.; ZOWE, J.: A unified theory of first and second order conditions for extremum problems in topological vector spaces. Mathematical Programming Study 19, 39 - 76 (1982)

[5] BLANCHARD, P.; BRÜNING, E.: Direkte Methoden der Variationsrechnung. Ein Lehrbuch. Wien-New York: Springer, 1982

[6] BLASCHKE, W.: Über den größten Kreis in einer konvexen Punktmenge. Jahresber. d.Deutschen Math.Vereinig. 23, 369-374 (1914)

[7] BLOECH, J.: Optimale Industriestandorte. Würzburg-Wien: Physica Verlag, 1970

[8] BLUM, E.; OETTLI, W.: Mathematische Optimierung. Grundlagen und Verfahren. Berlin-Heidelberg-New York: Springer, 1975

[9] BLUMENTHAL, L.M.; WAHLIN, G.E.: On the spherical surface of smallest radius enclosing a bounded subset of n-dimensional Euclidean space. Bull.Amer. Math. Soc. 47, 771-777 (1941)

[10] BROKATE, M.; KRABNER, P.: Some remarks on multiplier rules in Banach spaces. Freie Universität Berlin, Fachbereich Mathematik. Preprint no. 84/79, 1979

[11] BRYSON, A.E.; HO, Y.U.: Applied Optimal Control. New York-London-Sydney-Toronto: J. Wiley, 1975

[12] CANTOR, M: Vorlesungen über Geschichte der Mathematik, 1. Band. Leipzig: B.G. Teubner, 1880

[13] CHENEY, E.W.: Introduction to Approximation Theory. New York: Mc Graw-Hill, 1966

[14] COLLATZ, L.; WETTERLING, W.: Optimierungsaufgaben (2. Auflage). Berlin-Heidelberg-New York: Springer, 1971

[15] COLONIUS, F.: A note on the existence of Lagrange Multiplipliers. Appl. Math. Optim. 10, 187-191 (1983)

[16] COURANT, R.; ROBBINS, H.: Was ist Mathematik? Zweite Auflage. Berlin-Heidelberg-New York: Springer, 1967

[17] COURT, N.A.: Fagnano's problem. Scripta Math. 17, 147-150 (1951)

[18] COXETER, H.S.M.: Introduction to Geometry. Second Edition. New York-Chichester-Brisbane-Toronto: Wiley, 1969

[19] DANTZIG, G.B.: Lineare Programmierung und Erweiterungen. Berlin-Heidelberg-New York: Springer, 1966

[20] DANTZIG, G.B.: Reminiscences about the origin of linear programming. In: Mathematical Programming. The State of the Art. Bonn 1982. (Eds. A. Bachem, M. Grötschel, B. Korte), 78-86. Berlin-Heidelberg-New York: Springer, 1983

[21] FAN, K.: Asymptotic cones and duality of linear relations In: Inequalities II (Ed.O. Shisha), 179-186, London: Academic Press, 1970

[22] FIACCO, A.V.; McCORMICK, G.P.: Nonlinear Programming: Sequential Unconstrained Minimization Techniques. New York: Wiley, 1968

[23] FIKE, C.T.: Starting approximations for square root calculation on IBM System/360. Comm. ACM 9, 297-298 (1966)

[24] FLEMING, W.H.; RISHEL, R.W.: Deterministic and Stochastic Optimal Control. New York-Heidelberg-Berlin: Springer,1975

[25] GALE, D.: The Theory of Linear Economic Models. New York-Toronto-London: Mc Graw-Hill, 1960

[26] GASS, S.I.: Linear Programming. Methods and Applications. Fourth Edition. New York-Toronto-London: Mc Graw-Hill,1975

[27] GIRSANOV, I.V.: Lectures on Mathematical Theory of Extremum Problems. Lecture Notes in Economics and Mathematical Systems 67. Berlin-Heidelberg-New York: Springer, 1972

[28] GOLDSTINE, H.H.: A History of Numerical Analysis from the 16th through the 19th Century. New-York-Heidelberg-Berlin Springer, 1977

[29] GOLDSTINE, H.H.: A History of the Calculus of Variations from the 17th through the 19th Century. New York-Heidelberg-Berlin: Springer, 1980

[30] GUSTAFSON, S.A.; KORTANEK, K.O.: Semi-infinite programming and Applications. In: Mathematical Programming. The State of the Art. Bonn 1982. (Eds. A. Bachem, M. Grötschel, B. Korte), 132-157. Berlin-Heidelberg-New York-Tokyo: Springer, 1983

[31] HADLEY, G.: Linear Programming. Reading: Addison-Wesley, 1962

[32] HESTENES, M.R.: Calculus of Variations and Optimal Control Theory. New York-London-Sydney: Wiley, 1966

[33] HESTENES, M.R.: Optimization Theory. The Finite Dimensional Case. New York-London-Sydney-Toronto: J. Wiley, 1975

[34] HEWITT, E.; STROMBERG, K.: Real and Abstract Analysis. Berlin-Heidelberg-New York: Springer, 1969

[35] HOFFMANN, K.H.; KORNSTAEDT, H.J.: Higher order necessary conditions in abstract mathematical programming. J.O.T.A. 26, 533-569 (1978)

[36] HOLMES, R.B.: Geometric Funktional Analysis and its Applications. New York-Heidelberg-Berlin: Springer, 1975

[37] IOFFE, A.D.; TICHOMIROV, V.M.: Theorie der Extremalaufgaben. Berlin: VEB Deutscher Verlag der Wissenschaften,1979

[38] JOHN, F.: Extremum problems with inequalities as side conditions. In: Studies and Essays: Courant Anniversary Volume (Eds. K.O. Friedrichs, O.E. Neugebauer, J.J. Stoker). New York: Interscience, 1948

[39] JUNG, H.E.W.: Über die kleinste Kugel, die eine räumliche Figur einschließt. J. Reine Ang. Math. 123, 241-257(1901)

[40] KARUSH, W.E.: Minima of Functions of Several Variables with Inequalities as Side Conditions. Master's Dissertation, University of Chicago, 1939

[41] KING, R.F.; PHILLIPS, D.L.: The logarithmic error and Newton's method for the square root. Comm.ACM 12, 87-88, (1969)

[42] KIRSCH, A.; WARTH, W.; WERNER, J.: Notwendige Optimalitätsbedingungen und ihre Anwendung. Lecture Notes in Economics and Mathematical Systems 152. Berlin-Heidelberg-New York: Springer, 1978

[43] KNOWLES, G.: An Introduction to Applied Optimal Control. New York-London-Toronto: Academic Press, 1981

[44] KÖTHE, G.: Topologische Lineare Räume I. Berlin-Heidelberg-New York: Springer, 1966

[45] KRABS, W.: Optimierung und Approximation. Stuttgart: Teubner, 1975

[46] KRABS, W.: Einführung in die lineare und nichtlineare Optimierung für Ingenieure. Leipzig: Teubner, 1983

[47] KUHN, H.W.: On a pair of dual nonlinear programs. In: Nonlinear Programming, Ed. J. Abadie. Amsterdam: North-Holland, 1967

[48] KUHN, H.W.; TUCKER, A.W.: Nonlinear programming. In: Proceedings, Second Berkeley Symposion on Mathematical Statistics and Probability. Berkeley: Univ. of California Press, 1951

[49] LEE, E.B.; MARKUS, L.: Foundations of Optimal Control Theory. New York-London-Sydney: J. Wiley, 1967

[50] LEMPIO, F.: Separation und Optimierung in linearen Räumen. Dissertation. Hamburg, 1971

[51] LEMPIO, F.: Tangentialmannigfaltigkeiten und Infinite Optimierung. Habilitationsschrift. Hamburg, 1972

[52] LINNEMANN, A.: Higher-order necessary conditions for infinite and semi-infinite optimization. J.O.T.A. 38, 483-511 (1982)

[53] LUENBERGER, D.G.: Optimization by Vector Space Methods. New York-London-Sydney-Toronto: J. Wiley, 1969

[54] LUENBERGER, D.G.: Introduction to Linear and Nonlinear Programming. Reading, Mass.: Addison-Wesley, 1973

[55] LYUSTERNIK, L.A.: Conditional extrema of functionals. Mat. Sb. 41, 390-401 (1934)

[56] LYUSTERNIK, L.A.; SOBOLEV, W.I.: Elemente der Funktionalanalysis. Berlin: Akademie Verlag, 1965

[57] MANGASARIAN, O.L.: Nonlinear Programming. New York: McGraw-Hill, 1969

[58] MARTI, J.T.: Konvexe Analysis. Basel-Stuttgart: Birkhäuser, 1977

[59] MAURER, H.; ZOWE, J.: First and second order necessary and sufficient optimality conditions for infinite-dimensional programming problems. Mathematical Programming 16, 98-110 (1979)

[60] MEINARDUS, G.; TAYLOR, G.D.: Optimal partitioning of Newton's method for calculationg roots. Math. Comp. 35, 1221-1230 (1980)

[61] MOURSUND, D.G.: Optimal starting values for Newton-Raphson calculation of $\sqrt{x}$. Comm. ACM 10, 430-432 (1967)

[62] NEUSTADT, L.: Optimization. Princeton, N.J.: Princeton University Press, 1976

[63] PENOT, J.P.: On regularity conditions in mathematical programming. Mathematical programming study 19, 167-199 (1982)

[64] PIERCE, J.G.; SCHUMITZKY, A.: Optimal impulsive control of compartment models, I: Qualitativ aspects. J.O.T.A. 18, 537 - 554 (1976)

[65] PONSTEIN, J.: Approaches to the Theory of Optimization. Cambridge: Cambridge University Press, 198o

[66] RADEMACHER, H.; TOEPLITZ, O.: The Enjoyment of Mathematics. Princeton, N.J.: Princeton University Press, 1957

[67] ROBINSON, S.M.: Stability theory for systems of inequalities, Part II: Differentiable nonlinear systems. SIAM J. Numer. Anal. 13, 497 - 513 (1976)

[68] ROCKAFELLAR, R.T.: Convex Analysis. Princeton, N.J.: Princeton University Press, 1970

[69] SMITH, D.R.: Variational Methods in Optimization. Englewood Cliffs, N.J.: Prentice Hall, 1974

[70] STEINHAGEN, P.: Über die größte Kugel in einer konvexen Punktmenge. Abh. Math. Sem. Hamb. Univ. 1, 15-16(1922)

[71] STOER, J.; WITZGALL, C.: Convexity and Optimization in Finite Dimensions I. Berlin-Heidelberg-New York: Springer, 1971

[72] STURM, R.: Maxima and Minima in der elementaren Geometrie. Leipzig: B.G. Teubner, 1910

[73] TOLLE, H.: Optimization Methods. Berlin-Heidelberg-New-York: Springer, 1975

[74] TICHOMIROV, V.M.: Grundprinzipien der Theorie der Extremalaufgaben. Leipzig: B.G. Teubner, 1982

[75] VAN SLYKE, R.M.; WETS, R.J.B.: A duality theory for abstract mathematical programs with applications to control theory. J. Math. Anal. Appl. 22, 679-706 (1968)

[76] VERBLUNSKY, S.: On the circumradius of a bounded set. J. London Math. Soc. 27, 5o5-5o7 (1952)

[77] WETS, R.: Grundlagen konvexer Optimierung. Lecture Notes in Economics and Mathematical Systems 137. Berlin-Heidelberg-New York: Springer, 1976

[78] WOLFE, P.: Explicit solution of an optimization problem. Mathematical Programming 2, 258-260 (1972).

[79] YOSIDA, K.: Functional Analysis. Third Edition. Berlin-Heidelberg-New York: Springer, 1971

[80] ZACHARIAS, M.: Elementargeometrie. In: Enzyklopädie der Mathematischen Wissenschaften III 1,6. Leipzig: B.G. Teubner, 1914-1931

[81] ZOWE, J.; KURCYUSZ, S.: Regularity and stability for the mathematical programming problem in Banach spaces. Appl. Math. Optim. 5, 49-62 (1979)

SYMBOL INDEX

SUBJECT INDEX